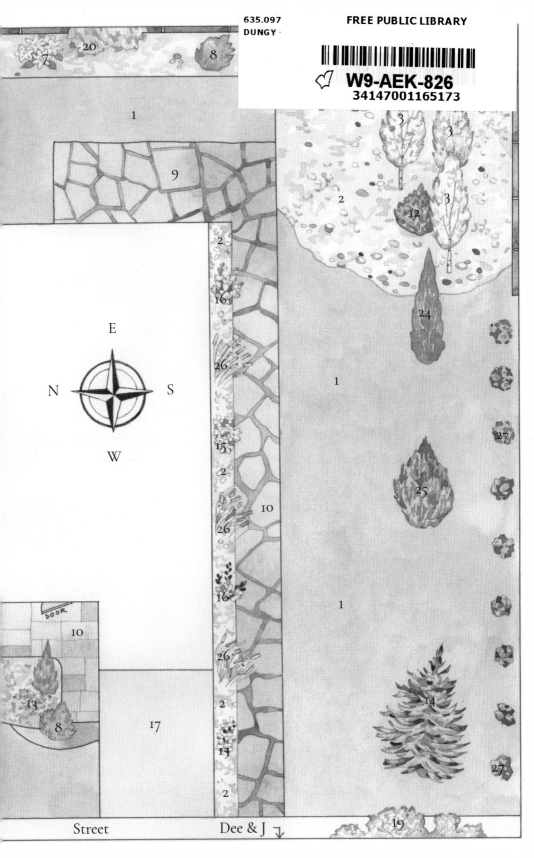

ALSO BY CAMILLE T. DUNGY

Essays

Guidebook to Relative Strangers: Journeys into Race,
Motherhood, and History

Poetry

Trophic Cascade

Smith Blue

Suck on the Marrow

What to Eat, What to Drink, What to Leave for Poison

Anthologies

Black Nature: Four Centuries of African American Nature Poetry

From the Fishouse: An Anthology of Poems that Sing, Rhyme,
Resound, Syncopate, Alliterate, and Just Plain Sound Great
(with Matt O'Donnell and Jeffrey Thomson)

Gathering Ground: A Reader Celebrating Cave Canem's First Decade
(as associate editor, with Toi Derricotte and Cornelius Eady)

Soil

The Story of

A BLACK MOTHER'S GARDEN

CAMILLE T. DUNGY

Simon & Schuster

NEW YORK   LONDON   TORONTO   SYDNEY   NEW DELHI

Simon & Schuster
1230 Avenue of the Americas
New York, NY 10020

First Simon & Schuster hardcover edition May 2023

SIMON & SCHUSTER and colophon are registered
trademarks of Simon & Schuster, Inc.

For information about special discounts for bulk purchases,
please contact Simon & Schuster Special Sales
at 1-866-506-1949 or business@simonandschuster.com.

The Simon & Schuster Speakers Bureau can bring authors to your live event.
For more information or to book an event, contact the
Simon & Schuster Speakers Bureau at 1-866-248-3049
or visit our website at www.simonspeakers.com.

Interior design by Ruth Lee-Mui

Manufactured in the United States of America

5   7   9   10   8   6   4

Library of Congress Cataloging-in-Publication Data has been applied for.

ISBN 978-1-9821-9530-4
ISBN 978-1-9821-9532-8 (ebook)

for my family—

Callie

Ray

my parents and sister

Aunt Mary and those gone before

plants and birds and beasts tame and wild—

for all of you

Camille holding native flax blossoms

bloom how you must i say
—LUCILLE CLIFTON

Hawthorn branch, with berries

I sat in my Realtor's Lexus while she toured me around Fort Collins, Colorado, during the waning weeks of May 2013. At that time of the year, I know now, flurries from cottonwoods fill the Northern Colorado sky. Fluff gathers in yards and along the road's margins like snowdrifts. Apple trees, mountain ash, hawthorns, chokecherries: all their white petals fall onto sidewalks, pile onto windshields. Neighborhood by neighborhood, petal-strewn curbs whizzed by the car windows—disorienting me. All that whiteness swirled around us, and I wondered what I was getting my family into. I thought that, by May, Colorado would be heading into summer, but the wide-flung whiteness made me think the cold of winter might be all this town could offer me.

My shoulder-length black locs and the Realtor's brunette waves swayed in and out of each other's side views. We never got out of the car. I didn't want to look closely at anything yet. Later that summer, Ray and I would come back and choose a house to buy. On this trip, I just wanted the lay of the land.

Fort Collins is a long hour's drive from the northern edge of Denver, and just thirty miles south of Wyoming's wide-open rangeland. The American plains spread behind us as we drove. Jagged granite peaks of the Rocky Mountains loomed ahead. I'd arranged the real estate tour because I planned to accept a job that would transplant my NYC born-and-raised husband, Ray, and our three-year-old daughter, Callie, from Oakland, California. This was a final chance to see if I felt safe moving my family to the state where I'd been born but where I hadn't lived since I was a toddler.

For more than 140 years, Fort Collins has grown around its university and old town center like rings of a tree. Every two decades or so, a new ring raises subdivisions from the surrounding farms and ranches and the remaining grassland prairies. The town's steady growth meant we could afford to own a house with a yard, leaving behind the rented apartment that overextended our budget in Oakland. The university offered me a position as a full professor, and a secure teaching position for Ray as well. The community's respect for education meant an accessible system of high-quality public schools for our daughter. All these possibilities excited me. I wanted a home where our family could stretch out and root down in peace.

But I had to figure out what was going on with that white mess all around me.

Was I going to suffer teary eyes and headaches thanks to all the shedding cottonwoods? Would my allergies flare in this environment? Were those fallen ash petals going to be a hassle to clean up? And what about the people in this predominately white town? Would they welcome our Black family? My mind worked as it wondered. I looked to understand this place's disposition based on the evidence of my interactions—and the history of others' interactions—with the living world.

Carrying with it seed looking to grow a new tree, cottonwood fluff can travel on the wind for up to twenty miles. As May unfolds into June here, the ash's white petals will finish. The hawthorn petals, the chokecherry buds, and the apple blossoms too. The trees will go green, leafed out for a new year. Gorgeous in their own ways. What fruit comes in place of the petals will welcome birds, who'll take berries for themselves and, also, to feed their young. Birds who set up nests sometimes, or sometimes just settle for a moment's rest. I have to stay in one place long enough to see it, but there is promise all over when I look.

At our tour's end that late-spring day, the Realtor asked me, "Where do think you want to live?"

I've been working to answer that question ever since.

Two stages of a firewheel, Gaillardia pulchella

One October morning six years after that tour in my Realtor's Lexus, a large white truck dumped seven cubic yards of shredded cedar mulch to cover our driveway. Then the truck dumped another seven cubic yards of compost-enriched topsoil onto the street in front of our house.

The Wednesday the soil and mulch arrived was not a calm day. Since the weekend, wind had blown at rates of up to forty miles an hour. Drastic weather changes here kick up a barometric fuss. The night before, our house shook in gusts more than once. Fallen leaves blew everywhere and gathered with dust motes and stray trash in crevices around our porch and patios and shrubs. Such winds come to Colorado to warn us that, though the day rose warm, by nightfall our yards could be blanketed in snow. That October Wednesday was not the ideal day to open our home to many hundred dollars' worth of soil.

"Why couldn't they deliver on Friday?" Ray asked a few days before the truck arrived.

"Because Wednesday is the day it is coming," I said.

Set on completing the first steps in our yard project before Halloween—a week away, when costumed and sugar-crazed kids would crawl through our neighborhood—I hadn't considered the wind. I simply called the landscaping supply company to arrange delivery. And the delivery was there.

Ray and I danced around the mound in the street, trying to protect our soil from the wind. Our friend Tim stopped by briefly on his way to deliver a jacket to his daughter at the nearby elementary school. He couldn't stay to help, but he watched our frenzied motion for a moment.

"If this isn't a metaphor for marriage," I told Tim, "I don't know what is."

The plan had been to convert the water-hogging south lawn into a drought-tolerant, pollinator-supporting flower field. But between a book tour that had me out of town nearly weekly from June to October, and the work of overseeing some interior renovations when I was home, most of 2019's planting window had slipped by me. A snowy spring had run right into the blazing heat of summer. Then it was fall, and soon the snow would start flying again.

"What are you going to do about the south lawn?" Ray asked several times that summer. He was enthusiastic about the idea of creating a yard with more breadth of life than the turf that had come with the house. Like me, he'd read climate reports the *Guardian* and the *New York Times* dropped into our email. We both knew the catastrophic impact on local pollinators wrought by decades of American suburban lawn culture. Knowing that sometimes even good ideas overwhelm me until I let them go, he warned, "We should get started before it's too late."

I had a quiver full of excuses: It had been a very hot summer, with temperatures often exceeding 100 degrees Fahrenheit; leaves wilted in the yard; flowers died in the dry heat; Fort Collins gets only about fifteen yearly inches of rainfall, compared with a national average of around thirty inches.

I told Ray I needed time to research low-water, semi-arid-climate-loving plants.

He gave me a look.

I need time to research was one way to say *I'm not going to do what you're asking right now.* On my most exhausted days, which were many, I figured not planting a large pollinator garden would be fine. No one used that patch of the yard but for a solitary rabbit.

"Camille." Ray wasn't giving up. "What do we need to do to get that pollinator garden started before winter?"

Ray and I had been married more than a decade by then. "Love is patient," said a reader at our wedding.

We married on the summer solstice. A date decided largely by venue availability, but one that feels like the most instructive kind of coincidence. The summer solstice is the longest day of the year. I love those expanded hours of brightness. But the night of the solstice, the hours when we and our wedding guests danced in June moonlight, is also when darkness starts to beat out the light. The day we announced our love also reminds us that there could be less of what we love from day to day.

During our years together, Ray learned that sometimes my big ideas need to be broken into manageable sections. So he took the wheel and talked to our lawn guy, a man I worried about offending with my plan to drastically reduce the amount of lawn we paid him to mow. "Andy's on board," Ray reported. "Explain what you want. He'll lend a hand."

The first week of October 2019, Andy arrived with a sod cutter. In less than forty-five minutes, he'd sheared several two-by-four-foot strips of sod, bundled them like enormous Fruit Roll-Ups—green on the inside—and loaded them into his trailer. Roots waved up from the rolls' dry, gray, exposed soil. Suddenly, we had 720 square feet of stripped earth to amend in preparation for planting come spring.

The reader at our wedding declared, "Love is kind."

For a week, I admired that weed- and grass-free stretch of compacted, reddish-gray clay. The air smelled dusty and expectant without the lawn's scent heavy on the breeze. I was grateful that Ray had engaged Andy, and thankful to both of them because they pushed the pollinator garden project into life.

Whether a plot in a yard or pots in a window, every politically engaged person should have a garden. By politically engaged, I mean everyone with a vested interest in the direction the people on this planet take in relationship to others. We should all take some time to plant life in the soil. Even when such planting isn't easy.

Our south plot's newly exposed surface was as hard and crumbly as the sloughed-off parts of an old red brick. Too dense for most plants' delicate roots, the clay seemed dead and unproductive, lacking the organic materials—the microbes and fungi and insects—that differentiate useless dirt from what gardeners call soil. The weight of the thick sod had compacted the earth so that water pooled on the surface instead of penetrating. I had to pound my shovel to break through. The clay crusted and cracked and formed rocks the size of Callie's nine-year-old fists.

I was going to have to amend that patch—to feed it, aerate it, enhance it—if I wanted to welcome new life. Even though this tract of land was accessible, I'd have to work to build a garden. I wanted to work to build the garden.

The green of growing things calms me. Plants stabilize me. And I am interested in the patience that is required as I wait for growth.

For the politically engaged person—any of us—such patience is a key to survival. Patience is a kindness that carries me through long days and longer nights.

In May 2019, weeks after I first hatched the plan to replace the south lawn with a pollinator garden, I gave a poetry reading in Manhattan as part of a venerable series. I was excited to accept the invitation, but the night became a disappointment before it even began.

I'd neglected to check Callie's school calendar before I agreed to the trip. This was the year her class participated in the annual third-grade egg drop on the last day of school. Each kid designed their own container in which to protect an egg that a beloved custodian would drop onto the playground's blacktop from the roof. An egg landing intact earned the child much-coveted bragging rights. Callie had been excited about this for months. But I was on a plane to New York City the morning of her big event.

Ray recorded the launch and happily successful landing of Callie's egg, but watching on a phone after the fact is not the same as being there.

The poetry reading promoted an anthology celebrating the varied voices of the United States. The evening's readers represented several races and ethnicities, a kind of attention to inclusivity I admired. But a few days before my flight, I found out that I was the roster's only woman. I brought this to the attention of the event coordinators, and they said it was too late to correct the lack of gender equity. As a concession, they said that I and the other readers should make a point of reading others' poems to that end.

When I joined the seven male readers at the venue, the organizer reminded us of our time limit and suggested I read first. As if to be first is to be anointed. *The first woman to do that!* or *The first*

Black person to do this! I hear such praise so often as a cause for celebration. But to be first is often lonely and vulnerable.

I read my poem from the anthology, as well as one poem each by two other women: a wry, pointed poem by Jane Mead and a focused, hopeful poem by Audre Lorde. I kept to the specified time limit. Then I sat down. Like an obedient girl.

Near the podium stood vasess filled with bright red gladiolas. Big and loud, stems tall as my torso. Gladiolas are summer flowers, tropical plants. It was late spring, still cold, that New York night. In what heated greenhouse or chemically drenched country would these trumpeting beauties have been raised and cut down? One of the reasons Ray and I wanted our own garden was to surround ourselves with beautiful flowers without participating in an industry that often sprays women and brown people with toxins for the sake of more profit and yield.

The men at the podium, every one, read over their times. They read their own poems from the anthology. Then they read others. Not *others* as in other people's—women's—poems, which was the idea conveyed to me. No. These men read other poems *of their own.* As the reading wore on, the hothouse gladiolas seemed more and more like reflections of the size and shade of my rage. I'd flown to New York on one of the biggest days in Callie's elementary school experience to read a single poem of my own and watch men drown out my voice and the voices of all the other women in the book. The red gladiolas rose from their huge vases. I had no way of knowing then that I'd missed Callie's last in-person closing day of elementary school.

After the reading, the men headed to a bar for a celebratory drink. I was not welcomed to join them. I knew several of these guys, have known them for a very long time. This was not the first nor the last time they sent me cues—a kind of concentrated cold-ness—that made it clear they did not want me in their space.

About a decade earlier, at a writer's conference I attended along-side several of the same men, I asked why they excluded me from that evening's plans when, in the past, we all spent hours laughing and talking together. One of the guys' lips curled into a sneer. "Be-cause you went and got married," he said.

"But you've been married this whole time we've known each other," I countered.

"That's different." Before turning his back on me to join the rest of the guys as they prepared to go wherever men like them go, he said, "You're a girl."

He said that word, *girl*, with such disdain I wondered how I ever managed to believe I'd been a valued member of that tight circle of men.

I woke up the morning after the New York reading feeling acutely alone. I missed my bed, my family, but I couldn't go home right away. I had scheduled a session for a new publicity photo. I have some headshots I love. Flattering ones taken by talented women photographers. A radiant one Ray took before he was my husband—when we were dating—when I didn't know he was fo-cused on me. Seeing the photo he captured, I immediately realized I should stick with a man who saw the best of me—who loved me that well—even when he knew I wasn't looking. But years had passed. More white and gray had moved into my locs. In my eyes, I saw a different kind of exhaustion than previous photos revealed. So my agent connected me with a photographer to capture a new set of headshots.

Much of the session took place in the photographer's Alphabet City studio. But when the thrill of being in a New York photo shoot waned, and my anger about the previous night returned to douse my inner light, he proposed a solution. I write so much about nature, he explained. It made sense to take some photos of me interacting with the natural world.

The photographer held a membership in a nearby community garden. We passed three vegetable gardens along the way to our destination. It made me happy to see them. In the process of making meals, it is important to sometimes get our own hands dirty. But vegetable patches require a quid pro quo between plants and humans, where the plants get to keep growing so long as they produce tasty, nutritious fruit for human consumption. The photographer had an ornamental garden in mind for me. We walked to a space more like the kind of garden Ray and I wanted to build. A garden that seemed to have nothing to offer humans but beauty.

Every person who finds herself constantly navigating political spaces—by which I mean every person who regularly finds herself demoralized and exhausted by the everyday patterns of life in America—should have access to such a garden.

There was a koi pond. Stands of lilies. Though the sun warmed us that May afternoon, it was a bit chilly. Still tightly poised, most green things were not yet ready for the showy stage of blooming. But my God! Even early in the season, this garden enchanted me.

I saw evidence of camaraderie and a collective sense of humor among the garden's members. A comically illustrated poster on the corkboard showed how gardening offered THE ORIGINAL FULL-BODY WORKOUT. Shoveling, hauling, digging, plugging, squatting, raking, lifting, and pruning all engaged different muscle groups. The garden welcomed merriment. Members had to work to enjoy the space, but they could enjoy the space while they worked.

Elephant ears the size of my chest! Catalpa leaves the size of my head! The smell of chlorophyll and loam, like the petrichor I love so much after fresh rain. In that garden, I breathed deeply and happily for the first time since I'd boarded the plane to New York.

The religious tradition I grew up in claims there was The Garden, where Adam and Eve lived in ease. And then, because of Eve's misdeed, there was the rest of the world. The wilderness into which they were cast. A world filled with labor and struggle and strife. I disagree. If the sweat and sore muscles of the original full-body workout yield spaces like the one where I stood that morning in Alphabet City, such a lot doesn't feel like a horrible fall from God's grace.

In the photo we took in the garden, I am standing in a bank of maple leaves. I am looking off somewhere, toward other beings who thrive, as I believed myself, in that moment, to also be thriving. My smile is genuine. Because in gardens I find hope.

As he packed up his equipment, the photographer told me the mayor of New York worked to get rid of gardens like the one where we stood. Often founded on vacant lots to reengage and resuscitate overlooked land, such gardens, according to the mayor, wasted potentially valuable property. Imagine the revenue that could be gleaned from that lot with a building full of shops and condos. Imagine how many people could be housed—and at what a high price—in that now-wasted space.

Such imaginings leave out the people who'd found home there already. The photographer and other members of the community garden. The koi and the songbirds and the butterflies I watched with excitement during the hour I spent there. In some cosmologies, worldviews I honor, these fish and birds and butterflies are also people. Living beings, with lives of value. The tree people who found space in that garden not afforded to trees on the average city street. If we felled them, what a high price we would pay. Those trees gave us a different kind of wealth. Carbon capture and a payout of oxygen. A space that absorbed the clang of surrounding streets and enveloped us in a dampened, cooling, calming quiet. A place to rest during the work of resistance, where my body lowered the cortisol spike I suffered the evening before. A place where I reveled in beauty. I can't quantify the economic value of beauty as compared with another human structure in what is now a garden. A garden is never wasted space.

My part of the prairie project, the name Ray and I settled on for the south yard's pollinator garden, involved calling the landscaping company to arrange delivery of amendments that accelerated the productivity of hard clay soil. After months of stalling, I made this call. Without consulting the weather report, or Ray.

For several years after we all moved to Colorado, my parents pointed out that I tended to say *my* garden, *my* bedroom, *my* house when I talked about our new home. None of these spaces should be considered mine alone, they suggested. "You should include Ray and Callie," they said. *Our* garden, *our* home. *Our* daughter, *our* life.

I try. But too often I slip back into thinking about what serves me alone. *I* was ready for the soil, and so *I* called for the soil. And so, one Wednesday morning late in October, a delivery truck with two enormous, improbably shiny white compartments filled with soil arrived.

I'd wanted to stand outside and watch, but the driver warned us about the dust. As a compromise, Callie, Ray, and I observed from the closed window of my (newly renovated) home office. I found the delivery thrilling. What engineering feats must go into figuring out how to lift that heavy bed and its otherwise secure tailgate so each payload, but only one payload at a time, poured into a controlled pile within feet of our front door. Mounds high and wide enough to fuel a dirt-biking child's dreams emerged out of a dust cloud, like figures in a magic show.

When I showed my student Jess a video of the truck dumping an SUV-size load of soil in what appeared to be the middle of the street, she observed my delight and laughed when Callie noted how admirably cars shifted course around the road's new obstacle. Next she noticed Ray emitting what might have been the world's longest, deepest, most exasperated sigh.

This, too, is a way of measuring love. How deep are our sighs? How do we learn to stand beside people whose actions and attentions we can't fully fathom, even as the world dumps a pile of crap and dirt and shredded promise just outside our door?

Normally at that time in the morning, Ray—who, like a male emperor penguin, is often the lead parent in our home—made sure Callie finished her breakfast, packed her lunch, and overcame her aversion to wearing socks. Then he'd hop on his bike and trail her to the elementary school before pedaling off to his university office. But, that day, during the windiest week of the fall, his wife had ordered a truckload of dirt. Ray rearranged his plans.

In less than ninety minutes, he had to lecture in front of his class of a hundred students. He still had papers to grade and a multimedia presentation to complete. Instead, he would spend thirty minutes driving to and from the nearest hardware store.

"Can you make sure Callie gets to school on time?" Ray asked after his long sigh. "I'm going to buy tarps."

I love this man.

"While you're there," I hollered as Ray started the car, "can you pick up whatever you call the kind of shovel that has a pointy end?"

In the immediate aftermath of receiving enough landscaping material to convert 720 square feet of what was formerly sod (and, honestly, a whole lot of dandelion, clover, crabgrass, creeping Charlie, and thistle) into an extensive native and naturalized flower garden, some people might hesitate to admit to their partner that they didn't know the name of a shovel with a pointy end. And it is possibly also unreasonable that a person about to undertake such a project would have used a small, bright-red, oval-ended shovel so vigorously that its head broke off in the cruel Colorado clay. To compound these failures, when I tried to repair the damage, I noticed, for the first time, the words CHILDREN'S TOOLS FOR WORK AND FOR PLAY inscribed on the broken shovel's back end.

I'm no landscaping expert.

I sometimes find it hard to believe anyone trusts me.

When Ray and I first started dating, I was in the initial stages of compiling the anthology *Black Nature: Four Centuries of African American Nature Poetry*, the first book to highlight Black America's long and varied legacy of environmentally engaged poetry. In those early days of the book project and my relationship with Ray, I flew from the Bay Area, where we both lived, to conduct research in his natal city.

With about half the number of people occupying roughly twice the square-mile grid, San Francisco is not nearly as crowded or built up with concrete and steel as New York. Lower Manhattan's densely packed buildings, bright taxis, and intimately pervasive buzz made me miss the man with whom I'd fallen in love. I called Ray while I walked on the tight, busy streets of the warehouse district near the old Poets House. "Where are you right now?" I asked.

Ray was on the fifth floor of the parking lot closest to our offices at San Francisco State University, where I was a professor in the Creative Writing Department and Ray taught as an adjunct in the Department of Africana Studies. The campus is only a three-minute drive from the Pacific Ocean, and the air was often heavy with salt-soaked fog.

"There's a duck on that car," said Ray.

"A duck?" Lake Merced bordered SF State's campus, but I had trouble believing a duck would fly all the way to the roof level of a parking garage.

"Yes. A duck. I mean, I think it's a duck." Ray stopped a bystander. "Is that a duck?"

The bystander told him the bird was a seagull. This made more sense to me, given the proximity to the sea and the amount of trash available for a scavenger bird on the campus.

There was nothing sublime about the way Ray or I interacted with nature at that moment.

I'd flown to New York City to research a book about Black poets who write about nature. But there I was, infatuated by the giant human footprint surrounding me and falling deeply in love with a New Yorker who couldn't tell the difference between a gull and a duck.

Ray said, "At least I knew it wasn't a pigeon."

I worried that I might be a greenwashed fraud.

At the time Ray misidentified the gull, we lived in a paved, stuccoed world. The only animals of note scavenged human waste. Six years after the duck incident, we moved here, to a planned Fort Collins community, where broad, green lawns can be as destructive to native environments as steel and concrete. But we were trying to build a more supportive environment for the plants, insects, birds, and small mammals who lived around us. The duck incident is one reason why—though neither of us quite knew what to do with the weather, the flora, or the unforgiving, hard clay soil outside our door—we committed ourselves to learn. The duck incident is one reason why, that October morning, we scrambled to secure the mulch and soil I ordered. I could have laughed at Ray that day on the phone call between New York and San Francisco. Both of us did laugh. We laugh still. Who confuses a duck with a gull? What I mean is I could have ridiculed him for what he didn't know. Instead, I realized in that moment how much I, also, do not know. About myself. About the world around me. How much I take for granted about the language I use to describe the world, and how much I could miss out on as a result of what I didn't work to learn. The duck incident showed us a new way to learn to love each other and to love the world in which we live.

Thanks to that duck, Ray and I became students of flying people. By the summer our daughter came into the world, three years after that transcontinental phone call, Ray and I owned more bird identification books than parenting books. We hiked together and subscribed to the Cornell Lab of Ornithology's email list. When we moved into the Colorado house, we installed bird feeders around the yard. By the fall of the soil delivery, we had five different types of feeders, supplied through an account at a store called Wild Birds Unlimited. Our plan for the prairie project included a birdbath with a circulating water supply and bee and butterfly hydration stations. We fell deeper in love with birds and bees and butterflies together. This work hasn't always been easy but, together, we developed a vested interest in our relationship with each other and with our on-the-wing neighbors.

When I told Ray I didn't know the proper name for the broken digging tool, he didn't laugh at me. He accepted the fact that I am still and always learning. This is part of how we say, "I love you. I love how you help me live in the world."

Just before Ray returned from the hardware store with two blue twenty-foot tarps and a shiny, adult-strength, pointed digging shovel, I managed to get Callie off to school. She even wore her socks. By asking the soil delivery man what size tarps he should buy, Ray made space for someone else's wisdom in our garden, which is one of the most important keys to survival and success we have learned. This is why the tarps he bought fit our needs.

Later that evening, I spent half an hour online researching the names of different types of digging tools and I shared the most informative article with Ray. We're in this pursuit together.

While we were engaged, Ray said we should have a very small wedding. Maybe just our parents and a few of our closest friends. That wasn't what happened. By some accounts, more of our wedding guests were our parents' friends than selected friends of our own. For this, we have reasons to be grateful.

During our wedding reception, the DJ invited all the married couples onto the dance floor. "Anyone married less than a day, sit down," he commanded. He started playing through a list of our favorite songs, a melodious score for something he called "the anniversary dance." As the newlyweds, Ray and I witnessed the longevity of the relationships of the people who had come to support us.

We married the week in 2008 when the state of California first issued same-sex marriage licenses, and so when, moments after we sat down, the DJ said, "Anyone married less than five days, please take your seats," a member of our wedding party and his new husband left the dance floor too. In the center of that couple's table, as on all the tables in the room, stood bouquets of the kinds of flowers you might find in a San Francisco Bay Area yard. Pink and yellow and cream and red and orange gerbera daisies. Delicate sprays of baby's breath. White-petaled, yellow-tongued calla lilies. Little white and yellow rosebuds. Bright, common, dependable flowers I love.

A month. Six months. A year. Two. Three. At each marker, more couples—Black, white, Asian, Latinx, Native, and mixed race—sat down. But at forty years, forty-five, even fifty-five, couples still smiled as they stepped in sync around the dance floor. The final couple danced all the way back to their table when the DJ called out fifty-eight years. The beads on her deep blue dress glittered in the light of the candles, and his brown eyes twinkled as he watched her move. As I watched two people whose deep love I'd witnessed since I was a child, I thought of a line from a Nikki Giovanni poem: "Black love is Black wealth."

The DJ told us he'd never seen anything like the length of our anniversary dance. My mother thinks he meant he'd never seen so many Black people in successful long-term relationships, or he'd never seen such a diversity of races and ages and sexual identities gathered, in loving celebration, in one room. Maybe that was it. Maybe it was something else. As Ray and I watched the collective motion persisting on the dance floor, we understood differently—better—what it would mean for us to work to love each other for all our lives. And we understood differently—better—that we'd entered a community intent on helping us.

.   .   .

When Ray returned from the hardware store that windy October morning, we pulled the tarps over the piles, secured them with some broken floor tiles left over from our bathroom renovation, re-maindered hickory flooring from our home office renovations, and two-by-fours we found in the garage. Then, with less than twenty minutes to spare before the African American history class he taught began, we got in our car, and Ray drove us to campus.

Tim texted around three p.m. On the way to pick up his young-est daughter from school, he saw that the wind had blown up sev-eral corners of our tarps. The blue coverings waved like wind socks in a gale.

Ray and I scrambled to rearrange our plans and raced home, hoping to keep the wind from claiming all our soil and mulch. To-gether, we shoveled into container after container some of the dirt that stretched beyond the pile's margins and farther into the path of traffic. The more we shoveled from the bottom of the street-side mound, the more the mound spilled back into the spots we shov-eled. "We've passed the angle of repose," I told Ray. Cars driving around us slowed only slightly but never threatened to run us down.

While Ray and I worked on saving our dirt, I kept thinking of what I said to Tim earlier that day: "If this isn't a metaphor for mar-riage, I don't know what is."

I remembered some railroad ties the house's former owners had piled in the southeast corner of the backyard, discarded landscaping relics from the 1990s. Ray brought the heavy wood beams around the house, and we wrestled them into position on the tarps.

"The bees better never sting me again," I told Ray. There was something in this for us, yes. But this was also about a larger vision.

We were discovering how to use less water, how to benefit more living beings, and how to build something more beautiful, more nurturing to our human and nonhuman neighbors, than what we'd inherited when we moved into this house. When I said I hoped the bees wouldn't sting me, I gave voice to my hope—our hope—that the bees appreciated our efforts.

"I don't envy you that job," said a ruddy-faced man who cruised past us in a green-and-white Mini Cooper. The neighbor's joking tone reminded us that our work was visible, maybe laughable, to others.

As Ray and I wrangled another railroad tie onto the tarped mound in the road, the Mini Cooper returned. I thought of how delighted Callie was when the Cat in the Hat tidied up the big mess toward the end of the Dr. Suess book. "Did you bring a *Cat in the Hat* cleanup machine?" I asked the driver.

"It's not much," he said. "I found a few things behind my wife's she-shed." He offered us a couple of cinder blocks, more railroad ties, and a few heavy rocks. "Good luck!"

And just as quickly as he'd come, he was gone.

Golden columbine in the prairie project

Our two-story house looks like a child's drawing of a house. All squares and triangles set in the center of a fifth-of-an-acre lot.

On a small concrete patio just outside the front door stand pots of annuals. Strawberry-looking globe amaranth. Bright ruffled marigolds. Troll-doll-haired celosia in neon red and orange and pink. Showy fuchsia, whose white and pink and purple bells surround fuchsia-colored pistils like the skirts and legs of Moulin Rouge dancing girls.

To the right of this patio is the start of a stretch of lawn leading to a wide juniper bush as tall as I am. The bush grows between our lot and our northside neighbors' driveway.

I use the word so often, I looked up what *neighbor* means: "n. 1. a person who lives near another. 2. a person or thing that is near another. 3. one's fellow humans." Proximity to another. A kind of fellowship. Humanity. By happenstance of alphabetical arrangement in my *Webster's Dictionary*, the word *neighbor* shares a column and close proximity with the words *negress, negritude, negrophile*, and *negrophobe*. These are the kinds of alignments and accumulations and accidents of proximity I notice.

We built a raised vegetable bed toward the eastern edge of the fence we share with our northern neighbors at the start of our sixth summer in this house. Out of tedious rock beds near those raised beds, and also in the northwest corner near the patio, I cleared small patches for some welcomed plants. Still, a bank of river rock extends along most of the length of our house's north exterior wall. River rock overrun with bindweed and western salsify (which looks like an overgrown dandelion) and plain old dandelions too. I've done little in that north-central part of the yard to give the eye entertainment. The area mostly consists of a sizable chunk of lawn, measuring about four by fifteen yards, that Callie asked me to preserve so she has a place to run around with her friends.

It was partly to fight the dandelions that flourished in the hardscape along our northwest wall that I tore out some of the rock in the spring of 2014. Two kinds of creeping Veronica now carpet the space with their pea-size low-growing leaves and lentil-size purple and white flowers. Pink and blue and purple bachelor's button as well as blue flax fill the little plot too. They each have blossoms the size of a dime. The bachelor's buttons sport several stems with a bright crown of rays atop each. If I deadhead spent flowers for new growth to come in, these bloom from mid-June through October. Some large part of gardening, like some large part of living, is figuring out what to cut and when.

Curving toward the patio near that northwest corner—in the sunny patches around a no-longer-diminutive dwarf mugo pine that welcomed us to the house, and which it took me several years to learn to properly trim—delicate red and pink Shirley poppies grow if I've prepared the plot successfully and the season is right. In June and July, the cleared space sports grape hyacinth with heads shaped like blackberries and petals the color of blueberries.

Deep purple and lavender and pink and sunset orange and buttercup yellow and creamy white snapdragons grew out of the rocks and weed barrier that covered this patch the August we first moved in. The snapdragons and the dwarf mugo pine were the only living plants in that bed.

Poppies, bachelor's buttons, and snapdragons are considered annuals in our part of the world, where winters come sometimes as early as September. They complete their entire life cycle in one growing season. While perennials send down taproots, investing for their own future, annuals gamble everything on their seeds, hoping future generations might prosper somewhere else. By August the bachelor's buttons will host as many dried seed heads as flowers. The Shirley poppies will flower no more.

Plenty of perennials also gamble on seed heads. The blue flax seedpods look like tan ball bearings up and down the plant's stalks. The seeds scatter over our yard's northwest corner. Year after year, even as fresh shoots sprout from old root crowns, I can count on new plants to populate the areas I open for them.

One May afternoon in our seventh year here, Callie and I spent a few hours thinning the north lawn's abundant swaths of dandelion heads. We each had a bucket, and the fun was to see who could fill hers the fastest. She was nine, and rather than consider weeding a thankless chore, I wanted Callie to enjoy this activity. We concentrated on the flowerheads, catching them before they went to fluff and seed that would spread through the yard.

Dandelions are the type of perennial that diversify investments between the roots and the seed heads. The May afternoon in 2020 when Callie and I weeded, I decided not to concentrate on the ways the plants' roots also spread. Instead, we focused on what we could see above the surface. A spent dandelion flower can send each of its seed filaments as far as sixty miles. It's astounding the distance some seeds can spread.

"Did you know people used to want dandelions in their yards?" I asked Callie.

"What?" she said. "Why!"

My daughter used to love these little flowers just the same as any other flowers. Maybe more, because they seemed to offer themselves up so abundantly. She wove chains of dandelions into bracelets and necklaces. She presented them in tiny bouquets as sincere displays of her most earnest love. When their filaments turned white and readied to spread, she picked dandelion seed heads and blew, believing whatever she wished might come true. But I'd absorbed the late twentieth-century landscaping preference that dandelions were undesirable. Without meaning to, I passed it on to her.

I wanted to redirect our thinking.

I popped a dandelion into my mouth—stem, leaf, and flower.

"Are you sure you can eat that?" Callie wondered why I would eat something I seemed to despise.

European immigrants in the seventeenth century brought dandelions to North America because they missed the bright, tasty plant from their homeland. One benefit of the broadleaf perennial is that their deep and broad root systems pull nutrients and minerals from parts of the soil other plants can't reach. The north side of our house slopes, so the ground suffers more runoff than other parts of the yard, leaving less topsoil and more compacted clay. I may have been trained to revile these plants, but dandelions help our yard in many ways. They find bare spots in the hard ground where grass roots can't penetrate and, opportunists that they are, and hard workers, they dig in, sending taproots as deep as three feet. Breaking up nonproductive dirt, they start to create healthy, receptive soil.

From deeper in the earth, dandelions pull up minerals like copper, calcium, iron, manganese, magnesium, phosphorus, potassium, and zinc. Through their broad leaves and multiple flowerheads, they spread these and vitamins K, A, C, and nearly all the Bs across strata more accessible to other plants and animals. Dandelions are one of the most nutritious plants in our yard. Some experts say their greens are more beneficial (albeit more bitter) than kale.

"I am sure you can eat them," I told Callie. I described dishes we could make: roasted dandelion root tea, dandelion pesto, dandelion green salads, dandelion head fritters, pickled dandelion bud capers, pizzas topped with dandelion leaves.

She wrinkled her nose and insisted the plants should go in the green bin for compost, not in her mouth.

Callie continued collecting flowerheads and the little buds that come before them, and so did I. She had a pair of scissors and delighted in giving so many dandelions a trim. As we worked contentedly near each other, our buckets filled and filled and filled.

"Do you think they used to do it this way in the olden days?" she asked.

I thought she was asking if people born years before her practiced our weeding method, which, since we weren't pulling up the plants' roots, was effective only as a short-term cosmetic fix to keep me from worrying about neighbors accusing us of bringing down property values with our weed-infested lawn.

"If you were an enslaved child, weeding might well have been one of your jobs," I told Callie. "But I imagine it would have been a great deal less fun."

She stayed quiet a minute, mulling over what I'd just said.

"But do you think they used to cut their lawns this way?" she eventually added, her eyes focused on the scissors in her hand.

By "they" she meant "people in the olden days" which, for her, could mean any time before 2010, when she was born. To me, "the olden days" usually meant sometime in the centuries before my own birth.

In the olden days, I told her, people used to cut lawns with scythes (curved, knifelike tools with long handles) and sickles (curved, knifelike tools with short ones). I told her that a Black man named John Albert Burr patented America's first nonclogging rotary mower in 1899. Burr was enslaved until his teen years. Even after that, he worked as a field hand, probably with a scythe and a sickle, until his innovations and inventions caught the attention of patrons who helped him leave the fields and attend school.

I do a lot of the work in our own yard. I enjoy the physical and mental demands of a robust garden. I like knowing that I helped to build this vibrant ecosystem with my own hands. We're not against paying people to help us with tasks I can't or don't want to take on, but we've tried to spare the expense of hired labor. Still, because we don't want to buy and store our own lawn mower, we hire Andy to mow our lawn every other week. Andy is a sweet, lanky guy, with tousled brown hair, who grew up in Fort Collins and built a business that allows him to work outside, in the place he loves. I pluck the dandelions only in late April and early May, before the grass has grown high enough for Andy to mow. No one forces me to work. I don't labor in the yard in bad weather, or when I'm sick, exhausted, or bored with the task. I *enjoy* working in the garden. The agency to choose to work in the yard, in *our* yard, belongs to us alone. I wanted Callie to understand the difference between her situation and the situation of children like John Albert Burr, born enslaved, and, even after Emancipation, subject to the mercy of white patrons. No person with control over our lives or our bodies or our money or our futures told my daughter or me what to do with our hands. We chose to weed the dandelions in this fun though ineffective manner. Plenty of people had, and still have, no choice about putting in countless hours of difficult, poorly remunerated labor.

"I don't think it's a coincidence," I told Callie, "that the man who invented one of the first lawn mowers was Black."

Callie looked up from her work again, this time with an inscrutable expression. Then she went back to pulling weed heads from the yard as quickly as she could. "I suppose that makes sense," she said.

In the end, she'd beat me at our game.

Sweet William gone to seed

My family moved into our Fort Collins home on a Wednesday in late August 2013.

Outside the house, the yard maintained a rigid order, homogeneously green and gray. A regimented line of five conically perfect four-foot-tall cedars flanked the east fence under the shade of the back neighbors' honey locust tree. The former owners of our house stationed another cedar along the front walkway. A white spruce stood sentry near the garage door, while the dwarf mugo pine held its post on the north corner of the front patio. One more white pine and a giant blue spruce, along with two more cedars—one six feet tall but round and bushy, and the other nearly twenty feet tall and a model of conical grace—provided vertical interest on the south side of the house. A few more severely trimmed junipers and three-foot tall wintercreeper bushes—shrubby foliated plants with oval green leaves augmented by splashes of white—grew around the yard. A substantially weed-free lawn and beds filled with one-inch granite rocks spread flat around these taller spots of greenery.

This nearly monochromatic order was the first thing I set out to change.

That August we moved in, I found canister after canister of herbicides and pesticides on the worktable in the garage. Very few pollinators braved the poisoned turf or weed killer–drenched rock beds. They flitted from one rare dandelion to the next, then buzzed away, forlorn.

The word *acre* used to indicate the amount of land a man and his ox could plow in a day. Now the word indicates a more specific measure. An acre is 4,840 square yards, about the size of three-quarters of a football field. Our whole lot, the house included, is only a fifth that size. Still, I spend whole days kneeling on my dark brown knees, pulling decorative rock and landscaping fabric from one five-square-foot bed or another. Every spring, before the heat comes on, and again in fall, before the cold really digs in, I rip out a new section of rock or replace a bit of lawn with a flower bed.

The rock in the hardscaped beds is piled four and five inches deep. In a day, I can fill six twenty-gallon containers, each with a carrying capacity of about 240 pounds. Twenty gallons of rock is lumpy, more difficult to contain than, say, twenty gallons of milk. I can't let the rock brim over the top, so I also fill a few five-gallon buckets, each of which can hold about eighty pounds of rock. I use a hand truck to carry these containers to the driveway so my friend Gillian can pick them up in her Land Rover. Its sturdy shocks sag under the load. Gillian uses the river rocks in her dog run, keeping the dogs and their waste out of her own native plant garden.

Gillian's skin is nearly the same shade as my own, a color I replicated as a child with the Burnt Sienna crayon in expanded Crayola boxes. We have similar gold-red undertones, and we both wear shoulder-length locs. Each of us has been mistaken for the other in this town. Over the years, as plants we've swapped take root in one another's soil, our yards have grown to resemble each other's as well. While we lift many pounds of rock into the bed of her mud-caked Land Rover, Gillian tells me stories of her work as a wildlife ecologist, studying the effects of climate change on communities in

Montana, Alaska, and farther out into the world. She collaborates with these communities, as she collaborates with me, devising positive and sustainable changes for our homes and the planet.

After removing the rock, there is still work to do. I find plasticized landscaping fabric, laid down to prevent the growth of undesired vegetation. Compacted by years of the weight of crumbled granite, dirt and clay coat both sides of the heavy material. It's difficult to grip. Some clay-caked rocks continue to weigh down the fabric. I use my whole body to tug the material out of the ground, sometimes ripping or cutting it free from large sheets still stuck under rock-filled sections I haven't yet cleared. I am drenched in sweat by the time a day is done.

Efforts to reduce natural diversity nearly always result in some form of depletion. What I find once I clear the beds of rock and fabric would be of little use to a garden. It is hard clay, packed so tightly that its impermeable solidity resembles the escarpments in the famous Red Rocks Park and Amphitheatre an hour south of us. To break up the clay and build more welcoming soil, I amend the newly revealed areas with compost magicked from kitchen scraps, cold barbeque ashes, the garden's deadheads, and fallen leaves. Our two tumbling compost barrels provide one to fill with waste we might have otherwise tossed into a landfill and one to leave alone so rank scraps can mellow into dark, crumbly, earthy-smelling compost. When I amend new beds, I also add topsoil hauled in from a landscaping supply store called Hageman Earth Cycle. I love the environmental vision inscribed in the name "Earth Cycle." The Earth—all we do and are here—is part of a cycle: a beginning that leads to an ending that leads to another beginning again. I love climbing Hageman's small hills of soil—the product of years and years of interconnected growth, decay, digestion, and excretion that promotes more growth. I shovel some into a wheelbarrow to haul home in our Honda. The full-bodied participation in promoting an ecologically vibrant landscape excites me.

I salvage the earthworms I find under the river rocks, tossing the wrigglers back into my newly enriched beds. If I hired a Bobcat to scoop the rocks, the work would go more quickly. What has taken me days and days over seven years could be accomplished in one morning with the help of such heavy machinery. But I work slowly, extracting and replanting desirable vegetation whose roots have grown into the landscaping fabric I tear away. It may take me twelve hours to prepare a three-by-three plot. Then I sow wildflower seed, perennial starts, tulip bulbs, and irises' gangly rhizomes. Sometimes we work as a family. Sometimes I work by myself. Within months of such sowing, we enjoy a riot of color where once there was only a hard, rust-gray expanse of rock.

Entering our second fall in the house, I covered the grass in the center of the front lawn and in a poorly irrigated corner near the driveway with layer on layer of discarded cardboard moving boxes, fresh kitchen scraps, topsoil, compost, newspaper, and mulch. Gardeners call this the lasagna method. It's a difficult process, even if it doesn't wear on my knees like the reclamation of the rock beds. But such layering can transform turf into rich soil without pulling up sod. By mid-May 2015, before hummingbirds passed through on the way to their summer mountain homes and after our local robins were already nesting, I turned the piles with a shovel and pitchfork, aerating the plot and revealing receptive loam in which I placed seeds.

Every year, I start a new bed around Halloween, when temperatures have cooled and we can expect our first snows and hard frosts. I must wait until nearly June before I see the blooms and bursts of color I labored so hard to welcome. The bachelor's buttons and poppies. Blue flax, black-eyed Susan, echinacea's pinky-purple cone-flowers. Columbine and hollyhocks can take two seasons to produce flowers. Changing our environment from homogeneous to diverse is rewarding. But the process can be slow.

Our garden regularly ruptures my sense of progress and process and time. There is the forward trajectory of days into months, sea-sons into years. June's tight rosebuds will lead to July's full-crowned blooms. Evident and irreversible change, straight forward as an arrow toward its mark. But there is revolution in the garden as well. And reversals. Months and seasons and days turning so far forward they bend backward. I stand in the past and in the future when I stand in the present of our garden. Just as with grief, neatly outlined stages double back and return well after or long before I expect them to appear or be over.

For two summers, the *Viola adunica* and *Viola nuttallii* I saved from Andy's mower blades disappeared from the bed where I'd transplanted them. But as I walked around the yard one morning in 2017, I noticed the penny-size purple and yellow wild violets lifting stalks three inches from the ground, as if I were kneeling again on the hot July 2014 day when I dripped sweat into fresh holes I'd dug for those violets with a clay-caked trowel.

Trouble spots and sites of beauty erupt here again and again and again. The garden reminds me I must be both vigilant and patient.

The neighbors' Norway maple wears the burnt-red leaves of au-tumn from the spring until they turn a deep, regal purple in the fall. In the tree's shade, clusters of fragrant blue blazes hyssop reach the height of their floral display in early September, whereas the white blossoms of the early-blooming cutleaf anemone disclose themselves in the middle of May. When the tulips faded and the purple allium unfurled from green sheaths like butterflies emerging from chrysalides, Ray admitted that before living here he never realized that each plant has its own cycle.

"I mean, it's not like just one thing grows here," he said. My husband is well past his fiftieth birthday, but he stood next to me with the embarrassed cheeks and eyes of a six-year-old child waiting to be scolded for not learning his lesson. "I know it's crazy that it's taken me this many years to realize that there are different seasons for different plants."

Ray wondered how he could have lived so long without understanding the patterns that occur around us. I could blame the number of years he lived in densely built-up cities, but so much of our world is manufactured. Many of us do not have a chance to see these cycles in action. We can buy strawberries in the store no matter the season. Now that our family grows them in the garden, though, Ray and Callie and I see how the small perennial bushes progress from green, to flower, to fruit, to gray-brown clumps of dormant stems, and back through that cycle again.

Like the strawberry bushes, our alliums and tulips and *Viola nuttallii* and sunflowers and cutleaf anemone all keep their own cycles and seasons. Ray felt the soft green sheath around the purple allium's flowerhead. "I've never had a chance to pay attention to these cycles before."

Plants go into soil at different times and come up in their own time. Sometimes they seem to exist all at once. Sometimes not at all. This, too, is the reality of the speed, the slowness, the wildness of time as it passes in the garden.

. . .

When we moved into our house, the covenant for our homeowners' association specified that the yard should be "well maintained." In those early years, a woman walked around the neighborhood with a ruler, measuring too-tall grass and what she considered unwieldy or weedy vegetation, reporting homeowners to the HOA board for review and possible censure. Once, we left a compost bin where it could be seen from the sidewalk. The HOA fined us $25 a day until we concealed the bin more thoroughly. A $25 fine may not sound like a lot, but in just a month such a penalty could add up to $775.

Banish the tall grasses. Banish the milkweed. Banish the front yard onion patch, the failing squash and melon trials. A "well maintained" lot should have fewer rangy wildflowers, or none. Certainly no gangly vegetable stalks. In 2012, the year before we moved to Fort Collins, city officials in Tulsa, Oklahoma, bulldozed nearly everything a Black woman named Denise Morrison grew in her front yard after someone reported her garden, full of edible and medicinal plants, for violating zoning ordinances. When Morrison contested the destruction, a federal court refused to hear her case. She had no recourse to recoup her losses. Rather than leave the dried stalks and seed heads on our sunflowers for birds to perch on and munch as they stock up for winter, stories like Morrison's urge me to pull all remnants of summer plants out of the ground. Keep nothing brown around the house. Nothing aesthetically unsavory.

My family is the only Black family on our block. In fact, we are one of the few Black families in our entire town, the fourth-largest city in the state of Colorado. By few, I mean that only 1.5 percent of people in this city of nearly 170,000 identify in US census records as

African American. By few, I mean that any time Ray heads out on the bike trails, someone recognizes him by name. Callie is the only Black girl with two Black parents in her 440-student elementary school. This might help clarify my resistance to the kind of suburban American monoculture that the woman in my neighborhood tried to promote via the HOA's yard maintenance code. A culture that—through laws and customs that amount to toxic actions and culturally constructed weeding—effectively maintains homogeneous spaces around American homes.

But I am lucky. The neighbors I speak to claim to be grateful we moved in, cultivating the most heterogeneous environment on our street, both with our presence as a Black family and with our landscaping decisions. Our HOA eliminated its rules against "nonstandard landscaping," and the town of Fort Collins actively works to help residents create landscapes that support native plants and insect populations and lower the strain on our precarious water supply. Bees love the flowers in our garden. Sunflowers shade low-growing plants at their bases. If, as the author Michael Pollan writes, "a lawn is nature under totalitarian rule," then my yard reveals a very different sort of possibility—one in which you never know exactly what or whom you'll find.

By the fourth summer in this house, we counted several species of bee, more than two dozen kinds of birds, and a slew of moths and butterflies, including monarchs—the migrating species that used to travel to their breeding ground in Michoacán, Mexico, in flocks of millions, though the population has been depleted by as much as 85 percent since 1996. We've planted milkweed, a native North American species on which the monarch's caterpillars are dependent, and all sorts of creatures flock to the perennials' blooms. In 2015 and 2016, I plugged handfuls of sunflower seeds into quarter-inch-deep holes around the yard. Though sunflowers are annuals with no permanent root structure or bulbs from which to spring again each new growing season, mine reseed themselves year after year. Each new generation of those big-headed sun lovers grows as high as thirteen feet, delighting many species of neighbor, humans included.

Brilliant goldfinches hang out near our Nyjer seed feeders, eating sunflower petals, stripping the yellow circle of bright ray florets on the outer edges until only the brown disk in the middle of the flower remains. I would prefer if the goldfinches didn't denude my sunflowers, but I planted the sunflowers as much for them as for me. I've learned that birds eat flower petals, not just the seeds that form in the flower's central disk. The brilliance of the gold in goldfinches' feathers is partially the result of the saturation of yellow in their diet. I have sown sunflower seeds in even more beds to continue to attract these beautiful birds. The sunflowers, *and* the birds who eat them, fill me with joy I could not have imagined when I first moved here.

"I love watching the goldfinches in your garden," a neighbor told me one afternoon.

"You don't get them in your yard?" I asked.

"Not like you do!"

Birds take up regular habitation around our place. The goldfinches and their cousins the plainer pine siskins. House finches with sweetheart red plumage painting their breasts and topping their heads. Black-capped chickadees—white baseball-size puffballs with gray wings and black necks and crowns. Red-breasted nuthatches, whose blue-gray, black, and white backs arrow toward narrow, pointed beaks that are perfect for grasping beetles, spiders, and ants. In addition to seeds from our other flowers and feeders, these birds eat sunflower seeds. Mating pairs of all these species and more choose spots in our trees and bushes to nest and raise their young. Robins hopping through the yard each spring snack on the worms I so carefully preserve when I dig up the old rock beds.

To control an aphid infestation that had compromised my rudbeckia plants—flowers commonly known as black-eyed Susan—I once released nine thousand ladybugs. The ladybugs came in three pint-size cartons that the local nursery kept in a glass-front refrigerator. So the leaves of our rudbeckia held the many droplets—like after a storm—I sprayed the stands with water. The ladybugs had plenty of places to hydrate as they prepared to feast on aphids. Then I laid the open cartons on the ground and watched as ladybug after ladybug exited by foot or wing. After that release, I saw an increase in creatures feeding on the nectar, pollen, and seedpods of the healthy black-eyed Susan. Swallowtails, painted ladies, and the occasional monarch butterfly settled down on these and other flowers in the yard. Each summer, they continue to pass through the garden.

When our aspens succumbed to the scale that struck many of the trees in our neighborhood one year, and Ray and I hired a crew to cut down two dead trees in our northeast corner, bugs and birds and squirrels still had plenty of places to congregate in our yard. Places that did not exist before I began the work of diversifying the garden.

To be honest, though, keeping up with all of this feels futile sometimes.

By May 2017, it seemed as if every dandelion in the county had planted roots in our yard. After four years of no herbicides, decades of residual toxins stopped being a hindrance in the soil. Dandelions found an opening they have taken advantage of every spring since. I spend at least three five-hour early-spring mornings with my old-fashioned weed remover, plunging the pointed blades around dandelion crowns, twisting the vise, and pulling out weeds at their root. Sometimes I simply pluck the flowers and seedheads, as I did that May afternoon with Callie. It occurs to me, doing this work, that the main reason people prefer homogeneity is that it often seems easier to maintain. A couple of applications of chemical herbicide and my yard would look neat as a magazine photo.

There's a man who comes by and fixes things around our house. His name is Mr. Comfort, and it fits him. His eyes always twinkle. His hands, sunspotted from years of outdoor work, are strong and gentle enough to set most broken things right. I deferred to him when it came time to decide how best to install the rain barrel I bought to divert runoff from a downspout. He installed irrigation systems in several of my reclaimed flower beds. He retiled our bathroom and painted the walls of my home office. He helped me select the best caulk to patch a hole between our bedroom window and the siding so wind wouldn't spill into our house. The year the dandelion population first exploded, Mr. Comfort offered to help with the weeds.

I have been tempted. Not just the spring he made the offer, but every year since. My flower beds are spectacular. But curly dock, purslane, bindweed, and black medic—whose tiny-leaved, hairy stalks creep out in many directions from its clumping center like a menacing spider—sprawl over the unimproved sections. They detract from the overall grandeur of our lot.

Maintaining a poison-free yard means revising some of my opinions about which plants I want around me and which I do not. I want to stay open to surprise, to stay open to lives that look and act in radically different ways than I am used to or comfortable with. If I welcome goldfinches, I want to get used to losing some of the sunflowers' bright yellow rays. But also—this is one of the key glories—by cultivating diversity, I learn things I never knew I should know. Neither river rocks nor turf grass is edible, but the dandelions, purslane, sunflowers, coneflowers, California poppies, black medic, and curly dock I either cultivate or tolerate all have nutritive value. The variety in my yard can provide sustenance in all kinds of ways.

Our first winter in this house, the killing frosts of October and November turned our world gray. There were long months without flowers or the evidence of the vibrant life I craved. February came. Then March. And then April. And because of all the rock and turf in our yard, there was nothing to look at but gray and more gray until May came and, with it, a little green.

As I planted bulbs and seeds, and as I put in perennial starts to root down in our soil, and as I swapped cuttings with friends, planting new plots in their honor, dividing and rearranging tubers and bulbs, cultivating the diversity in my garden, my mood brightened earlier and earlier each spring.

By March 2020, a few weeks into the pandemic, I was already outside some mornings, watching for the doll-cup-size purple grails of pasqueflowers—*Anemone pulsatilla*—to emerge from Colorado's snow-covered ground.

Showy milkweed with a western honey bee

let grow more winter fat   wine-cup   western wild rose

so little open prairie left   little waves of bluestem   little

fuzzy tongue penstemon   quieter the golden currant
nodding onion quieter now as well

only a few clusters of Colorado butterfly plant still yawn into the night

where there once was prairie

a few remaining fireflies abstract themselves
over roads and concrete paths

prairie wants to stretch full out again and sigh—

purple prairie clover   prairie zinnia
prairie dropseed nodding into solidago
bee balm brushing rabbitbrush—prairie wants   prairie wants   prairie wants

Late-season rabbitbrush

Callie sang as she bounced on the giant blue exercise ball I'd used as a desk chair since I was pregnant with her.

> *La la la! Walking through the woods.*
> *Nobody to think about but me!*
> *Me me me. Nobody but me!*
> *All I have to think about is me.*

After the deadly and dreary weeks of late March and early April 2020—weeks that kept us inside the house with only each other for company—sharing her bright pleasure was a relief.

Weeks before, on a Friday in mid-March, as concern over COVID-19 spread and Colorado's governor shut down much of the state, Callie and I canceled a planned trip to drive eight hours south, to Durango, Mesa Verde, and the Four Corners. Our school district closed buildings for spring break and never opened them again that school year. Callie was in the fourth grade, and she was stuck at home studying the Utah and New Mexico Territories with her nearly fifty-year-old mother. The new version of her school day frustrated her, and the frustration made her sad. She was angry about what COVID and the shutdowns took from her. She communicated these complicated emotions by rebelling against the mother who directed her life in such a dissatisfying way.

But, for the breadth of her song, Callie appeared completely self-actualized. She bounced on my exercise ball, singing like she'd opened a treasure trove.

I sat straighter on the exercise ball than in a chair. I rotated my hips to activate muscles in my lower back and thighs. Sitting on the exercise ball when I wrote, I reduced the pain I suffered while writing. But in those early days of April 2020, Callie claimed the ball as her own. Though the ball was about two-thirds her height, she managed to haul it over the piles of books and papers on my office floor, out of the door, and down to the house's main level, where she rolled, bounced, and balanced for hours at a time. The exercise ball provided action, distraction. She took it out of my office so often I started using a standing desk. *Let her have the ball*, I thought, *if she wants it that badly.*

Callie bounced and bounced at the kitchen table that served as her workspace. The incessant bounces punctuating sentences she read off a remote-learning exercise glowing on the laptop before her. She read in a voice stretched thin. Not because she strained to make sense of the words, but because she found the reading excruciatingly dull.

She wouldn't focus on the social studies exercise long enough to complete her assignment. I drew a compass rose next to the US map on her worksheet, to help her differentiate east from west and north from south. Understanding that Colorado is west of Missouri seemed simple, but I still look for the mountains looming close in the west to orient myself before I flip the irrigation switch marked SOUTH LAWN. Though I knew Callie's problem was boredom, I wondered if the problem had more to do with an inherited flawed sense of direction than her inability to focus.

She'd been excited for our trip. Ray needed to stay home working, so it would have been a mother-daughter adventure. We'd planned to drive to the southernmost counties of Colorado so she could directly witness some of Colorado's most remarkable history and geography. I wanted her to experience the borderless expanse of the vast southwestern desert, where you can look out from the top of a mesa and see hundreds of miles into a landscape that's been divided only on paper. Standing inside such vastness, I feel as if I belong anywhere and everywhere at once. At Four Corners, Callie could have placed one hand in Colorado and one hand in Utah while planting her feet in New Mexico and Arizona—all parts of the former Utah and New Mexico Territories described on her worksheet. Maybe the problem was lack of context. An inability to connect with the words on her assignment. "There are only a few more questions to finish," I said. "You can do this."

To counter my own exasperation about the circumstances that saddled me with teaching my resistant daughter, I'd walked two laps around the outside of our house early that morning. At latitude 40.5° north, you could draw a fairly straight line around a globe from Fort Collins to New York City and on to Madrid. It would seem that cities at the same latitude should experience roughly the same weather conditions, but New York's elevation is just thirty feet above sea level, Madrid's is 2,150 feet, while Fort Collins's official altitude is 5,003 feet. Temperatures drop by about 5.5 degrees Fahrenheit for every thousand feet of altitude, assuming no clouds, wind, snow, or other variables. When it's 65 degrees in New York City and 55 degrees in Madrid, it's still only about 40 degrees in Fort Collins. April is often the snowiest month in Northern Colorado, and nighttime temperatures can fall below freezing until mid-May. But the world was rousing that morning. An iris leafed out of the mulch we piled onto the prairie project before the first heavy snowstorm the year before. In the backyard, where the morning sun shines the strongest, crocuses bloomed along the east fence. If I was the kind of person who could spend leisurely mornings sipping tea at a little wrought iron table in the garden, I'm sure I would notice even more. But by the time Callie started bouncing madly on the exercise ball, I'd been inside three hours, stuck in the kitchen's homeschooling space.

"Sit down," I told Callie for maybe the twenty-fourth time that morning.

I looked out the kitchen nook's bay window. A house finch flew down, to perch on a feeder. The finch's beak reached into the opening to pull out sunflower chips, dried cherries, and small bits of peanuts from a mix the bird food store calls a "no-mess blend."

Callie bounced up on the exercise ball, as if preparing to rocket away.

"Sit down!" I snapped. The words sounded vicious.

She sat. She tried not to cry.

As a kid in the 1970s and '80s, I heard a refrain constantly on sitcoms, in movies, in the neighboring houses whose rooms I walked through: "I didn't ask to be born!" Kids hurled the phrase to hurt the one who manifested their original pain.

I wonder if the ubiquity of that phrase in my childhood has to do with the increasing number of mothers entering the workforce during those years. The frustration women felt, trying to be breadwinners and still play the roles of mothers, pretty wives, bakers, cooks, house cleaners, gardeners, tutors, seamstresses, and PTA board members. There weren't enough hours. It was impossible to be the kind of mother those working women, their children, and, it seemed, the world, thought ideal. Perhaps what happened in those years was happening in my house too. I was a woman, and a mother, balancing competing expectations of what I should be doing with my "one wild and precious life."

Those words end the poem "The Summer Day" by Mary Oliver. Part invitation, part command: "Tell me, what is it you plan to do/with your one wild and precious life?" I have the power, the lines suggest, to decide what I do with my time. Though "The Summer Day" was first published the same year I graduated from high school, it feels as if I've known its final sentence my whole life. I've seen it stenciled onto more inspirational boards than I can number.

A 2013 *New York Times* article about Mary Oliver described her as "the kind of old-fashioned poet who walks the woods most days, accompanied by dog and notepad." Much of my education groomed me to expect such behavior from someone who writes about nature. "All day," says the speaker of Oliver's poem, she "kneel[ed] down in the grass," "idle and blessed." She kept still so long a grasshopper ate "sugar out of [her] hand." The poem's speaker asks, "what else should I have done?"

I'll tell you what I would have done: The dishes. The laundry. A pile of work for pay.

Forget feeding a grasshopper. At least three times a day, I'm figuring out what to feed my family.

Even if I bring my daughter along "to kneel down in the grass," being her mother means letting go of the idle.

Callie gave up on the Utah and New Mexico Territory worksheet for a while, turning to an assignment that she liked better. She recited the twenty-five prime numbers between one and one hundred, while I listened.

Mothers today pay hawkish attention to our children. The phrase "hawkish attention" is meant as an insult to mothers like me, who worry over our offspring, hovering and coddling and spoiling our kids. According to a Pew Research Center report, American mothers spent one and a half times more hours on kids in 2016 than mothers in 1965 did. The report also said that these early-twenty-first-century mothers spent three times more hours working outside the house. How could mothers nearly double the time spent caring for children while tripling the amount of salaried work they completed? I have some explanations.

Benefiting from improved appliances and wrinkle-free fabrics, ignoring unmade beds and piles of unfolded laundry, feeding their families from the microwave or restaurants, twenty-first-century mothers spend significantly less time on housekeeping duties than did the 1965 cohort. It is wholly unsurprising that Instant Pots and air fryers were two of the biggest appliance fads of 2019 and 2020. Without requiring much effort from busy cooks, both promise great meals in minutes. Part of the reason I've selected so many Colorado native plants for my garden: I need plants I can leave untended for weeks at a time while I take care of other things that demand my attention. I can't afford the time it takes to care for finicky species that struggle in Colorado's climate.

When she finished identifying the prime numbers, Callie organized them into combinations that added up to one hundred. The exercise had something to do with seeking efficiency.

Other reasons early-twenty-first-century mothers have more time to devote to childcare as well as work outside the home: we sleep less, we socialize less with our partners, and we spend less time with our peers. The last woman under the age of sixty I heard talk about playing gin rummy or bridge did not have children. When

Michelle Obama and her husband, the then-president of the United States, instituted a regular date night, the weekly activity made national news. Partly because of the expense of their security detail, but partly because many were shocked that the Obamas left behind work and their children to spend time alone as a couple. Some of the backlash was mitigated by the fact that Michelle's mother—oh, and a whole Secret Service detail—could be counted on at the White House to look after the kids. Criminalized for leaving children on their own, mothers in the early decades of this new century are under vast cultural pressures to stay in the presence of their children. I've read cases of mothers, Black, Latinx, and white, being jailed or cited by Child Protective Services because they let their children walk to school alone, or play in a park by themselves, or play *in their own backyards* without them. This is how we became hawkish.

Watching my daughter's continuous insistence on motion as she bounced on the exercise ball, I realized that many of those mothers who'd been reminded that their children "didn't ask to be born!" must have been as torn as I was. Here is the contract the culture has made: in becoming a mother, the one precious life I am expected to think about is seldom my own.

Thoreau was right about this much: "To be in company, even with the best, is soon wearisome and dissipating."

I am weary always and dissipated most of the time.

But this is not what I thought I'd be writing.

In the proposal for the fellowship that bought me the time to write a new book, I made it clear that I wanted to write about my yard. Those crocuses. The patches of purple iris. A cluster of

Mexican sunflowers, *Tithonia*, that volunteered in our back garden in 2018. At first, I suspected the *Tithonia* were an undesirable weed. Then they blossomed into flowers so gorgeous I brought friends to the house to see them. Nine feet tall and topped with orange blooms like a cross between a zinnia and a gerbera daisy, the *Tithonia* were wonderfully unexpected and perfectly suited to the back northeast corner plot where they grew. I wanted to spend a year thinking about the soil that surrounded me: what grew up from it, and why.

But even before I found myself overseeing my daughter's remote education during the pandemic, I was troubled by the seeming improbability of a Black mother writing about nature.

Before 2020 became *2020*, I reread *Pilgrim at Tinker Creek*—a book that, for myriad justifiable reasons, has remained a favorite for many.

"Oh! I love that book!" one woman exclaimed when I told her what I'd been reading. I'd traveled to the university where the woman taught. We were on our way to one of her classes.

"Have you read it recently?" I asked.

"I guess I haven't."

"You might want to." I wondered if her response would shift on returning to the book.

Over the course of a year, Annie Dillard, the book's author, takes long walks beside the creek that runs near her home. That is basically the plot. Dillard shares much in common with the speaker in "The Summer Day." The authors, Dillard and Oliver, both white women, grew up just over one hundred miles from each other—one in Pittsburgh, the other outside Cleveland—in similar communities. They're just ten years apart in age and were raised in a world that expected a degree of solitary grandeur from people who write about nature. As in "The Summer Day," Dillard spent many quiet hours on a "stroll through the fields." Walks along Tinker Creek led to lengthy reveries about perception and philosophy and something akin to monastic ecstasy.

When I told my colleague I reread Dillard's book, I actually meant I *listened* to an audio version while completing household chores. This is a way to feed my hunger for reading while also completing the work that needs to be done. While folding laundry, sweeping the kitchen, cooking the evening meal, I listened to Dillard's book, feeling as if I were off in a field of unmown grass somewhere in Virginia. Dillard's language is sublime, and I often got caught up in its reveries. When a panicked fluttering erupted in my own chest—anxiety that came on precipitously one afternoon as I helped Callie navigate her own frustrations and doubt—I correlated my body's physiological responses with a grade-school trauma Dillard describes early in her book. I absorbed Dillard's experience completely. The highest commendation for a writer I know.

Dillard's book—which like me, is nearing its fiftieth birthday— has been popular for so long writers use it as shorthand when they profess a desire to undertake the type of project demanding they stay close to home and look very closely at their surroundings.

Back in 2018, I had lunch with an editor in Manhattan. I felt so fancy drinking a noontime glass of chardonnay with a New York editor.

"What are you writing next?" she asked.

"I want to write my *Pilgrim at Tinker Creek*."

The editor looked at me skeptically, as if assessing the magnitude of both my ego and my aptitude. Her bright face briefly dimmed ashen with doubt before arranging into a more professionally supportive posture, and she said, "Sounds like you're asking a lot of yourself."

．　．　．

From my journal, February 4, 2020: "You must find your own quiet center of your life, and write from that."

This quote comes from a 1908 letter the writer Sarah Orne Jewett sent to one of my favorite novelists, Willa Cather. After which, such gems from Cather's pen! She wrote a novel nearly every eighteen months from 1912 through 1931, and she published still more beyond that. Some of my favorite depictions of the American plains and the West: *O Pioneers!* (1913), *The Song of the Lark* (1915), *My Ántonia* (1918). I particularly love *The Professor's House* (1925)—with "Tom Outland's Story," that exquisite novella within the novel, a morally complicated rendering of a white man's encounter with and excavation of an ancient cliff dwelling that sounds a lot like Mesa Verde, down to the European businessman who swooped in and crated all the ancient Indigenous American artifacts back with him to Europe. How grateful I am to Cather for her patient, unflinching depictions of landscapes I love and the people who live here.

I have been reading Cather since high school, returning when I need to recenter myself in the world. During the first year of Callie's life, late at night while I nursed her and tried to resolve myself to myself, I reread *My Ántonia* on my backlit phone. I needed something of how Cather's narrator, Jim, describes the girl Ántonia, how the western landscape shapes Ántonia. I felt grounded by the book's rendering of a woman coming into herself.

At some point in the future, maybe I'll write about what it means that so many of Cather's narrators are men. Or the fact that neither Cather nor Jewett married. Or why neither of them, nor Mary Oliver, nor Annie Dillard at the time she wrote *Pilgrim at Tinker Creek*, had or raised children. What paths their lives took instead. There are so many reasons women do and do not marry, why we may or may not have children. Too many for me to write easily about this aspect of any woman's life but my own. For now, I simply want to say that though many aspects of my life are substantially different from Jewett's and Cather's, I am grateful for whatever role Jewett's advice played in setting Cather's pen into motion.

But I worry. What if there can be no quiet at my life's center, or anywhere near?

While Callie worked on her assignment, I worked on steadying my breath.

Twenty minutes after I snapped at her for bouncing so much, I turned toward her and took her hands. "Do you know why I got so upset?"

"Because you could be writing your book, but because of this virus you are homeschooling me instead."

It is possible that, years from now, it will be clear I erred in my decision to be transparent with such a young girl about my feelings and what triggers them. But I don't want to lie to my own child.

"That's part of it," I said. "But this situation is not in any way your fault. I do not begrudge you the fact that schools are closed."

Sometimes, when I use words she hasn't heard before, she stops to ask what they mean. Other times, like when I trot out words like *begrudge*, she furrows her eyebrows and seems to give up on me.

"You did not ask to be here," I continued. "I invited you into this world. It's my job to help you learn how to navigate it the best way you are able."

She brightened a little. But only a little. She likes to be at the center of my world, but she does not like the feeling of carrying the weight of that world on her shoulders.

"Really," I told her, "you may well be helping me. Remember the song we made up that night at dinner?"

In the early months of 2020, I got riled up by a book. Not Dillard's *Pilgrim at Tinker Creek* this time. Some other book that fits into a pattern in nature writing that confounds and annoys me. As I set the dinner table, I slammed our plates down with such frustrated force that Callie asked what I was listening to on my headphones.

What infuriated me, I told her, was that the audiobook I'd just finished, like so many foundational environmentally focused books, seemed to have no other people in it. The (nearly always white) men and women who claim to be models for how to truly experience the natural world always seemed to do so in solitude. Just one guy—so often a guy—with no evidence of family or anyone to worry about but himself.

Beside the remote school station at our kitchen table, Callie bounced on the exercise ball again, remembering the tune she invented the night we first discussed this serial omission. Her brow cleared. Her eyes shone, and not with tears. "La la la. Walking through the woods!" she sang, moving her arms as if taking a hearty stroll. "Nobody to think about but me!"

In that moment, Callie and I entered the full measure of her joy. We sang different versions of her little song a few times. Even recorded ourselves for posterity. I texted the voice memo to some writing friends, Suzanne and Kate. Since the lockdown began, the three of us texted daily. In many ways their lives felt quite different from mine. They lived in vastly different landscapes. Both were white women in long-term relationships with white men. Neither had children. Both could go trail running or hiking for hours without packing snacks for several other people. Still we built a community of women writers adjusting to our new reality together.

"Me me me! Nothing but me," Callie and I sang in the recording I sent Kate and Suzanne. "Not a soul to think about but *meeeeeeeeee!*"

With the singing over, Callie settled down at the table to complete her cliff dwellers worksheet. I wanted to head upstairs to my study and find "Tom Outland's Story"—to get lost in Cather's lucid descriptions of a Southwest filled with intoxicating landscapes and the conflicting needs of different people. But it was time to eat. I started making lunch instead.

Callie carefully handles the thorny Hawthorn tree

I was stooped over in the laundry room, loading the dryer, when Ray asked, "What's the plan for dinner?" The light from the window cast a shadow of the side yard's white pine across my back.

In the summer of 2019, *Poets & Writers* magazine published an article I wrote that offered ideas about how to make time for writing despite a busy life. My advice felt quaintly plucky as I gripped a handful of wet socks. COVID-19 stripped many parents of the scraps of idle time I suggested writers use for themselves. Write during aimless minutes in the doctor's waiting room, I urged. Write during a child's dance lessons!

I opened that article with a quote from the Puerto Rican American poet Judith Ortiz Cofer: "I have to steal time from myself." So many narratives of women fighting their ways into art look like women stealing time. I think of Toni Morrison, mother of two, scribbling on the subway during commutes between her home and her office. Or my favorite poet, Lucille Clifton, whose brevity matched the length of her children's naps. Some of her poems are as short as seven lines.

Between 1965 and 1969, Clifton had six children under the age of ten. In 1959, when the poet was twenty-three years old, Clifton's own mother died. She "had no model," Clifton wrote. She had to figure out on her own how to balance mothering with the requirements of nurturing herself. During the late 1960s, Clifton starts a poem with these lines:

> Dear Mama,
>   here are the poems
>   you never wrote
>   here are the plants
>   you never grew . . .

Even early in her career, Clifton thought about the creativity some women never manage to cultivate. She wrestled with the responsibility she felt—perhaps the honor—to find a way to write poems despite obstacles. Her pen grew what one dictionary calls an "arrangement of living material"—a garden.

But if they value their art, women writers, especially women of color, and most especially mothers, must steal their own time to grow such gardens. Heists like these are not easily executed.

Even wealthy white women—for whom space for self-actualization has been carved out of the lives and hours and art of women of color and poorer white women—struggle to create their own art and gain the respect of power brokers in the art world. Where are the foundational stories of this entitlement—the fully focused life of the artist—being offered to women who place their families as a priority as central to their lives as their artistic achievement? Maybe I didn't see mothers in the canon of environmental literature because it has long been impossible for mothers to write narratives of a world where they can wander alone in the open, pausing long enough to let grasshoppers eat sugar from their hands. Maybe I don't see mothers in the canon of environmental literature because it's impossible for most mothers to create a world where they have nobody to think of but themselves.

Greensboro, North Carolina, where I attended graduate school in the 1990s, has many white picket fences to hem in private yards. Businesses shut doors and shuttered windows around six every evening. I imagine this was so the men who worked in those offices could make it home in time for the suppers that awaited them. One of the most famous poems by Randall Jarrell, my graduate school's most famous writer, is called "Next Day"—a poem written in the voice of a weary woman grocery shopper whose most vibrant years are as lost to her as long-since-exploded soap bubbles.

Adding detergent to the washing machine in our cramped laundry room in Colorado, I thought maybe the poem I first read all those years ago drew an accurate portrait. "Wait just a while!" I yelled so Ray could hear me over the noise of the dryer. "I'll start making dinner soon."

Randall Jarrell served a two-year term as the eleventh consultant in poetry to the Library of Congress, a position we now refer to as the US poet laureate. Since 1937, the office has been held by one venerated American poet or another. Jarrell served right before Robert Frost's turn in the position. Until 2015, when the Chicano writer Juan Felipe Herrera stepped into the role, all but one of the men so honored were white, Jarrell included. Jarrell, the model consultant in poetry to the Library of Congress. Jarrell, the plain-speaking poet from Nashville, Tennessee. Jarrell, the literary critic who showed no mercy for work he thought unworthy. Jarrell, who taught at what was then a women's college and wrote, says the critic Jonathan Galassi, poems about women who "are unable to define precisely or to respond creatively to their predicament." Jarrell, man-about-town. No woman poet from the region received even a scrap of the praise my program heaped on Randall Jarrell.

During the years I attended graduate school in North Carolina, Maya Angelou lived one town over. When called on to recite work that moved them, girls across the country selected her poetry. I brought one of her poems into class for a discussion about compelling contemporary writing. The professor looked over glasses that slipped toward burst blood vessels on his nose. "I don't trust the voice here. Who talks like that?" He regarded the photocopied page of Angelou's poetry I offered as if he wanted to lift it from my hand and toss it in the wastebasket. "That is not the kind of work we study here."

The poem Angelou wrote for President Bill Clinton's 1993 inauguration came up once during a conversation in the department's mail room. One of my professors slicked back his wavy hair and pronounced the poem "too pedestrian." Another professor asked, "Where's the music?" Angelou's fault, I understood, lay in her voice being too plain and too Black and too womanish for these men to connect with, and, at the same time, too connected to too many (of the wrong kinds of) people.

All my professors were white. Most were men. I never had a Black writing instructor until I was already a tenure-track professor myself and I used a professional development grant to attend a summer writing workshop that featured Black faculty. Straight out of college when I arrived in Greensboro at twenty-two, I was so young. I want to tell you I rebelled against the canonical concessions my education fed me. I want to tell you I read all I could. That I worked double-time to build a world that looked, at least on paper, like one I could believe in, one that would make serious readers believe in me. But this took years. I grew weary pushing against my program's predominant current.

On crisp days in late April 2020, red-, black-, and yellow-centered flowers with snow-white fluted petals opened all over the prairie project we planted in the south yard. I bought the bulbs purely for their name: the poet's daffodil. Their sweet, solid scent, a bonus reward that kept me coming back to pass time with the flowers.

Though I wouldn't have explained it this way then, Mary Cassatt's work grounded me during those years I attended grad school in North Carolina the same way the daffodils that bloomed during that pandemic-altered spring did. In the days I lived in Greensboro, I often drove to Washington, DC, to spend weekends strolling through galleries and museums. I visited *Woman Bathing* by Mary Cassatt nearly every time.

The woman in *Woman Bathing* is neither very young nor particularly old. We see her from the back, her striped dress folded down to her waist. Bent slightly over a washbasin, she cups bright blue water in hands that create ripples on the surface of the water. The print is both static and active. The woman depicted seems almost motionless, but I anticipate the moment she brings water to her face. A fragment of her slippered foot seems to point out from her skirt, as if to stabilize her body. Her dress is rolled down to the top of her hips, and the crease in her back is pronounced. The image reveals her entire back, but her breasts are indicated only by a slight swell extending beyond her armpit. Of the woman's face, Cassatt shows only her left ear and a hint of the edge of an eye. She is faceless and frontless. Cassatt liked to use mirrors to extend the space in her paintings. Here, though, the mirror reveals only a mop of hair and an abstracted scribble of flesh where the woman's face might appear. The subject is naked to the waist, but *Woman Bathing* captures a moment of intense privacy.

Thanks in no small part to her wealth—"Born to a prominent Pennsylvania family," the Smithsonian American Art Museum's biography begins—Cassatt traveled to Europe, where she staked a space for herself in the art world. On viewing *Woman Bathing* at Cassatt's first solo exhibition in 1891, Edgar Degas is said to have proclaimed, "I do not admit that a woman can draw like that." But draw like that she did. And Degas knew this to be true. He saw Cassatt's work back in the 1874 Paris Salon and declared, "Voilà! There is someone who feels as I do!"

By invitation from Degas, Cassatt joined the impressionists in 1877. She was the only woman, and the only American, to regularly exhibit work with Degas, Renoir, Pissarro, Monet, and Manet. A curatorial note from the Smithsonian American Art Museum highlights something that helped Cassatt stand out from that fellowship of artists—emphasis on the masculine bias in the word *fellow*ship—and part of what drew me to Cassatt's *Woman Bathing*:

> It is noteworthy that both Cassatt and Degas preferred to call themselves "Independents" rather than "Impressionists"; both always insisted on the integrity of form in their painting, whereas Monet, Pissarro, and others tended to dissolve form into light. Like them, she initially employed a high-keyed palette applied in small touches of contrasting colors. However, over time, Cassatt's style became less painterly, the forms more solidly monumental and placed within clear linear contours.

The solidity of the body in *Woman Bathing*, in comparison to its potential dissolution, seems to be part of the point.

During the three years I lived in North Carolina, I befriended an Irish artist who, along with her husband, taught on the faculty at UNC Greensboro. The jobs offered their family financial stability. Helen and Paul and their daughter owned a white frame house with deep green trim down the street from the subdivided Edwardian where I lived. I remember long talks with Helen, in the bright expanse of her back garden, about my adoration for Cassatt's work. Helen's own art is nonrepresentational, but she never maligned Cassatt or even hinted that I was naive to revere the figure-focused painter so deeply. Instead, she praised the squiggles and messy corners in Cassatt's bodies and lines. When I think of Cassatt, I think also of Helen's back garden—always boisterous to the point of looking nearly out of control. Flowers grew higgledy-piggledy. Tall among short, finicky cultivars alongside near weeds, in such abundance random disorder circled all the way back to irrefutable scheme.

Shut out by gender and class from the raucous cafés and public spaces frequented by her male peers, Cassatt drew a world from the enclosed gardens and parlors and opera loges of her life. Though she never married or had children of her own, her paintings often center the domestic—mothers and their children, women in their private hours. As inspiration and models, she used her family, relatives of her housekeeper, small children from the nearby village: *Lydia* [the artist's sister] *Crocheting in the Garden at Marly* (1880); a woman with a baby and another young girl (*The Caress*, 1902); *Mother about to Wash Her Sleepy Child* (1880); *Young Mother Sewing*, with a child leaning on her knee (1900); one of Lydia leaning forward on her arms while seated in a theater's loge (1879); a painting of a woman (Lydia again) in her finery, including gloves and a hat, which Cassatt called *The Cup of Tea* (ca. 1880–81). A woman portraying the world of women.

Those domestic spaces did not isolate Cassatt from men's

judgment. I remember my shock—not of surprise, but of recognition—when I read a note from Adelyn D. Breeskin, a former curator at the Smithsonian, stating that while advising her about a painting Cassatt called *Little Girl in a Blue Armchair*, "Degas actually went so far as to paint one area of the background himself." The man's hand reaching even into the private space of the girl's parlor.

I want to think of Cassatt as a fully self-actualized artist. A painter who was well respected, at least in Europe, at her time. After visiting Cassatt in 1895, Violet Paget, an author who went by the pen name Vernon Lee, called her "a self-recognizing artist." I love this. As if to be an artist I must first look into a reflecting pool and see my own face.

By the second half of the twentieth century, collectors and art historians considered Cassatt to be "America's greatest female artist." I borrow that phrase from J. Carter Brown, who wrote the foreword to the catalog for a 1970 Mary Cassatt exhibit at the National Gallery of Art. I want to believe Cassatt found some secret that I could also find if I work hard enough. But this desire is complicated by facts.

For the 1893 World's Columbian Exposition in Chicago, Cassatt produced what some art historians consider a masterpiece of modernist mural art: a fifty-eight-foot-by-twelve-foot allegorical triptych commissioned for the Women's Building on the subject of the modern woman. In the left panel, "Young Girls Pursuing Fame," three young women reach toward the back of a cherubic naked child who flies out of reach. Some of the girls seem on the verge of flight themselves as they run and leap in pursuit of Fame. Even the ducks at the left edge of the panel seem about to fly. The world conspires *with* the girls in their pursuit. In the mural's massive center panel, "Young Women Plucking the Fruits of Knowledge or Science," women and girls of all ages hold fruit from a bountiful tree. One woman climbs a ladder and hands fruit to a girl. Another woman, likely dressed in white, pulls fruit from a branch.

I say *likely* dressed in white because no reliable record of what colors Cassatt used exists. For reasons not entirely clear, Cassatt's mural is lost. The only visual witness we have left are black-and-white photographs taken in 1893.

Nearly at the center of the mural, a woman who looks to be in her forties or fifties stands holding a fruit-filled basket. Her face and body are turned slightly. She stares in the distant direction of the right panel, "Arts, Music, Dancing." In this panel, one woman strums a long-necked stringed instrument, while a second dances and a third woman, wearing an intricately detailed floral-patterned dress, looks on.

Almost fifty when she created this monumental work in the garden studio behind her rented château in northwestern France, Cassatt had a trench dug that she could fit the enormous canvas inside. Because she didn't want to endure his reproach, she never let Degas see the work. In a letter to her friend Louisine Havemeyer, an art collector, Cassatt wrote:

> I am going to do a decoration for the Chicago Exhibition. When the committee offered it to me to do, at first I was horrified, but gradually I began to think it would be great fun to do something I had never done before and as the bare idea of such a thing put Degas into a rage and he did not spare every criticism he could think of, I got my spirit up and said I would not give up the idea for anything.

She completed the mural and shipped it to Chicago, where it was installed near the ceiling of the exhibition hall. Visitors had to crane their necks to see it.

Cassatt received $3,000 for the painting, the equivalent of $86,000 in 2020. Enough to recoup the expenses she accrued creating the mural. The same amount paid to male artists with similar works in the exhibition. Women in pursuit of fame indeed. Think of those girls Cassatt painted in the center panel, sharing the weight of a basket loaded with the fruits of knowledge!

My friend Helen collected the flowers from her Greensboro garden into bouquets. Big vases filled with huge-headed peonies, black-eyed Susan, lavender. In baggy linen pants, twine for a belt, she walked through the yard bending with scissors when she saw a bloom that suited the rowdy composition of the vase—deadheading spent blossoms all the while. Helen told me she didn't change her last name when she and Paul married because women artists too frequently disappear inside the names of the men to whom they are legally bound.

The exposed flesh of the woman bent over her washbasin in *Woman Bathing* looks as if it could be cast in Grecian marble. Her back, arms, and neck are nearly the same color as the washbasin, the mirror's frame, and the washstand that fill much of the canvas. My understanding of French suggests that the title in that language, *La Toilette*, foregrounds bathing itself—the acts of washing and dressing. Though it is the color of the crayon called "flesh" in my childhood, the woman's body seems less the object of the artist's gaze. What's important is the act of the woman coming into herself.

I think of the print, even now, when I am alone in the shower. One of the few times I can be in the quiet center of my own life.

The day before April's Utah and New Mexico Territories worksheet drama, it was my turn for tears. Thanks to a man who came to Colorado more than a century before and made a mint on silver, I was supposed to be enjoying my Guggenheim year. Fellowships endowed by the vast Guggenheim fortune have distributed $400 million to eighteen thousand artists and scholars since 1925. In 2019, after twenty years of applying, I won one of those fellowships. For the first time since Callie was born, I could afford to spend eight hours a day writing a book.

I hung a framed letter in my office during the first week of 2020, part of which read:

> *I HEREBY CERTIFY, That Ms. Camille T. Dungy, Poet, Fort Collins, Colorado, Professor, Department of English, Colorado State University, has been appointed by the Trustees of the John Simon Guggenheim Memorial Foundation to a Fellowship for the period from January 1, 2020 to December 31, 2020.*
>
> *During this period, Ms. Dungy will devote herself to Poetry.*

Printed on creamy paper that felt like linen, with an embossed seal and the hand-inked, stylized signature of the foundation's vice president, the letter officially and "respectfully recommended" that "all persons" to whom it is presented treat me with "esteem, confidence, and friendly consideration." I was to dedicate the year to capital-P poetry! That capital letter underscoring the gravitas and honor my writing deserved.

But things did not work out that way.

Coronavirus disease 2019 first appeared in Wuhan, China, in late 2019. Before Americans even really started talking about the virus, Ray had already lost an aunt who lived in the Caribbean. By mid-March 2020, several US cities, counties, and states, including Colorado, issued official stay-at-home orders. The virus had killed nearly sixty thousand Americans by late April—more than fifteen thousand in New York State alone. On May 24, the *New York Times* published brief obituaries memorializing one thousand of the one hundred thousand Americans lost to COVID thus far. Though Ray's uncle Joseph lived in the city all his life and died a casualty of the disease's disproportionate toll on Black and Latinx communities, his name didn't show up in the *New York Times*. The third immediate family member to contract the virus in early 2020, Joseph's wife, spent more than a week in a New York ICU before she moved to a rehabilitation facility—still, thankfully, though, in the end, all too briefly—alive.

Even after the deaths in Ray's family, the precautions taken by our own small family in Colorado seemed to be about securing some greater good. I figured that, if we followed protocols, the virus wouldn't harm me directly.

But the impact of the pandemic took a toll for which I was not prepared.

We thought Callie's remote schooling would last a few weeks, maybe a month. Then the school board announced that classes wouldn't convene in person again until sometime in the fall of 2020. Neither Cambridge University in England nor Stanford University in California planned to resume in-person classes until fall 2021. Stanford's date then moved to January 2022. Some institutions pushed in-person start dates back even further. Unlike me, Ray still had to teach classes. In a year already slated to be one of the most demanding of his academic career, he had four days to retool his entire syllabus and teaching style so he could instruct eighty students online. Ray locked himself in his home office. He worked all day and until three and four most mornings. He's a good father, attentive. He helped when he could. But he mostly had to leave us to take care of ourselves. When Callie's classes shifted online that March, her teachers shared curriculum materials to guide learning. But they couldn't do much more than that. It fell to me to administer most of her lessons.

The late-March evening before I yelled at Callie about bouncing too much at the kitchen table, I cried about the situation COVID put me in. Ugly, unflattering tears. Gasps and sobs that threatened to wake Callie. "I've lost my writing time to this virus!"

"They closed all of Broadway!" Ray said. "The New York City Ballet has canceled its entire spring schedule!"

How many of my fellow artists were out of work completely, Ray wanted to know. How many were left without a way to pay for food, rent, or medical care? Ray wanted to know why I thought I was so special that pain and death and hard times wouldn't touch me.

When Ray looked at me incredulously, I curled into a ball and clenched my fists in grief and rage. "I have a right to mourn the loss of this time!"

Ray remained unmoved.

He was furious, really. What was I mourning? A promise I made up for myself when I hung that letter on the wall? My temperament infuriated him.

I couldn't understand his anger. Mine was an attitude I cultivated as the natural temperament of the self-recognizing artist. I was supposed to devote the year to capital-P Poetry. I had every right to be frustrated that the demands of the household kept me from my art.

"You will never be satisfied with life as it really is," he said.

Ray needed me to understand that people were dying. Businesses permanently closed. Why wasn't I grateful to still be alive, housed, and fed? We were all doing what we could. We were lucky.

Someone needed to take care of our daughter and, of the two of us, I had the time. Ray needed me to stop crying about some precious, pretentious, solitary writing hours.

"What's this book called again?" asked a friend.

When I asked writers and teachers and book-loving friends if they'd read the book about the creek walker lately, initially I wanted to know if they thought Annie Dillard's long 1970s reveries translated to our early-twenty-first-century sense of time. Would a contemporary reader have the patience for the fifty-year-old book's multipage musing on the preening and protective ways of a common but elusive muskrat?

"I can't imagine anyone being interested in that book long enough to finish it," said my friend.

"Oh! But it's a seminal text," I insisted

According to the *Webster's Dictionary* I've kept near my desk since May 1991—a high school graduation gift from my one Black female K–12 teacher—my language did some work for me: "Seminal (adj.). 1: pertaining to, containing, or consisting of semen. 2: highly original and influencing the development of events: *a seminal artist: seminal ideas.*" The language I used when I insisted on the book's fundamental importance directly connected it, and its woman author, to the effluence of matchless men.

In *Pilgrim at Tinker Creek*, Dillard seemed to seek this alignment. She wrote gorgeous descriptions of the world, but she seemed to just walk and look and think metaphysical thoughts all day. She appears as an individual genius. I kept wondering where her people were. Did she never wash clothes? Did she ever argue—or do anything at all—with her husband?

As COVID tensions rose in America in March 2020, I texted Kate and Suzanne, the eventual recipients of Callie's and my "walking through the woods" song. I asked them, "Why doesn't anyone in foundational environmental literature seem to have to do the dishes?"

"There's so little quotidian honesty in nature writing during that generation," Kate responded. "I have to wonder if it's because publishers felt like there wasn't a tolerance for it."

Dillard herself said she "didn't think anyone would want to read a memoir by 'a Virginia housewife.'" In a 2015 article in *The Atlantic*, Diana Saverin wrote, "She left her domestic life out of the book—and turned her surroundings into a wilderness." But I want to ask Dillard why she didn't take the chance. Writing as a Virginia housewife, she might have inspired some 1970s readers to say, "Voilà! There is someone who feels as I do."

Dillard was twenty-seven when she wrote *Pilgrim at Tinker Creek*—only five years older than I was when the famous white men of the South and their antipathy toward art that did not reflect their own lives drove me to Mary Cassatt. We were alone—the two of us, so many of us—unseen in the wild dream landscapes of the famous white men.

We went outside our homes to find answers to the questions at the center of our lives.

Once outside, Dillard celebrated the isolation that sent her there. *La la la. Walking through the woods*, her pages seem to sing, *not a soul to think about but me.*

The performance feels intentional, like Dillard plugged her ears and screamed, *La la la. I'm not listening!* In the years just before she wrote *Pilgrim at Tinker Creek*, white-controlled public schools in Virginia actively resisted federal mandates requiring the enrollment of

Black children. One nearby standoff led to the 1968 Supreme Court case *Green v. County School Board of New Kent County*, which some civil rights historians say desegregated America's public schools far more expediently than 1954's *Brown v. Board of Education of Topeka*. The *Green* decision impelled schools in Roanoke, through which the Tinker Creek of Dillard's book runs, to fully integrate by 1970. That same year Cecelia Long became the first Black person to graduate from Roanoke's Hollins College, integrating the institution from which Dillard herself graduated in 1967 (with a BA) and 1968 (with an MA). The writer's husband, Richard Dillard, taught at Hollins. The couple lived just off campus. But Annie Dillard mentions none of these worldly details in her book about the world.

Though the turmoil of integration roiled immediately around her, Dillard maintained a willful segregation of focus and care. Borrowing some of Dillard's own language for a 2005 article he wrote in praise of her "impish spirit," fellow solitary-wood-walker Robert Macfarlane said, "In 1971 . . . Dillard moved to Tinker Creek, a valley in the Blue Ridge Mountains of Virginia. She lived alone there for four seasons, in a house 'clamped to the side' of the valley, 'facing', as she put it, 'the stream of light pouring down.' "

Continuing our text thread, Suzanne wrote, "I reread Thoreau and Abbey recently and I couldn't believe how misanthropic they were, especially male-centered and racist. Aren't we also a part of nature?"

I took Suzanne's last question to mean two things. Aren't humans also animals—not set *apart from* but, rather, *a part of* the natural world? *And* an insistence on asserting the importance of those of us who are so often erased from or maligned in books held up as environmental masterpieces of the nineteenth and twentieth centuries.

The conversation with Kate and Suzanne rang a bright bell

within me. Reading these so-called seminal texts, I feel excluded. The authors' inability to see me means that I have trouble picturing myself in the worlds they depict.

But I do exist.

Instead of accepting erasure, I learn to write a story for myself.

In 2009, I edited, wrote for, and published *Black Nature: Four Centuries of African American Nature Poetry*. One of the most remarkable statements the anthology made was that Black people write with an empathetic eye toward the natural world. Just as strong as the pull of legacies of trauma this nation inflicted—and inflicts—on Black people, some of us are pulled toward stories of hope and renewal. We're at peace upon the land. But because of erasures from so many narratives about the great outdoors, the idea that Black people can write out of a deep connection to nature—and have done so since before the founding of this nation—comes to many people as a shock. Conducting a literature review of poems included in the key nature poetry anthologies and journals published between 1930 and 2006, I found only six poems by Black poets. That's all. In eighty years of the environmental canon.

The editor of *Orion*, Sumanth Prabhaker, implemented a similar review of that magazine's pages. His count revealed that, between 1984 and 2018, writing by white men monopolized 78 percent of a publication that claimed to celebrate "Nature, Culture, Place." By expanding the range of writers and artists included on its contents page, Prabhaker and the editorial staff at *Orion* help to change the way readers understand and connect with "culture" and "nature" and "place." Content by white men—that global minority with outsize representation in so many spaces of power—is down to 25 percent.

Women, nonbinary people, and writers of color now have significantly more space in the magazine.

"Thank you," one Black poet said when I requested poems for *Black Nature* back in 2007. "I have been writing this way my entire life, but no one has ever seen me in this light."

To systemically exclude the lives of your neighbors from the space of your imagination requires a willful denial of nearly every experience outside your own. *La la la. I'm not listening.*

It's not just her Black neighbors' civil rights struggles that Dillard erased from *Pilgrim at Tinker Creek*. Such books erase just about everyone. "Dillard adopts the whole 'man-alone-in-the-wilderness (or in her case the pastoral)' trope," Suzanne added to our thread. "I mean, Edward Abbey was generally with one of his four wives out there in the desert, but they never show up. It's pure fantasy."

"That's part of why I like Amy Irvine's *Trespass* so much," wrote Kate. "She's so f-ing honest."

Published in 2018, Irvine's next book after *Trespass* is the contrarian meditation *Desert Cabal*. That book grapples with the implications of the fiftieth anniversary of the publication of Edward Abbey's seminal—in this case, there could be no more appropriate descriptor—*Desert Solitaire*. The same Abbey who stayed in the desert as frequently by himself as with one of his wives and their children.

Why disappear the people who people your world?

"Irvine doesn't get that level of love for *Trespass*," I wrote to my friends. "Partially because she was so busy raising her kid that she couldn't do the promotional work. But partly because nature dudes like a certain kind of story."

The nice thing about texting is I can carry on a conversation while vacuuming or stirring risotto. Sometimes mere seconds pass between messages. Sometimes long days. The flow of thoughts can meander as well. "It's noteworthy that Dillard wanted to write in a 'genderless' way (read: presumably male)," I eventually continued, "but they wouldn't let her publish the book as A. Dillard."

Suzanne, who had been offline, came back to tell us some of the phrases an editor used to replace the simple dialogue tag "I said" in her manuscript. Many of the substitutions magnified a kind of servile femininity: *I plead, I confess, I admit, I bustle, I apologize.* "As in," Suzanne wrote, " 'I'm sorry,' I apologized."

"Every time the word *bustle* comes up," wrote Kate, "we're doing a shot of tequila."

"I need to read *Trespass*," wrote Suzanne.

Though *Trespass* doesn't include descriptions of tequila shots, in *Desert Cabal*, Irvine writes about the precautions she takes as a woman. Including giving careful thought to what she drinks and with whom. She writes about the ways her life—the life of a white, culturally and economically privileged woman—is often in jeopardy in the wild. Especially when she's in the company of men. I tried to text that I found Amy Irvine's writing "fan-freaking-tastic." But my phone autocorrected to "Irvine is fan-freaking-tasty."

"Siri has already started in on the tequila," wrote Kate.

I returned to an earlier moment in our conversation—"There's so little quotidian honesty in nature writing during that generation"— and started my own contrarian list. Several books by contemporary writers do pay significantly more attention to the realities of domestic life than I've seen in previous generations. For *The Book of Delights*, Ross Gay wrote small daily essays about things that delighted him in this terrible, dangerous, everyday world. At some point in *The Book of Delights*, Gay even describes someone doing the dishes. And I know of at least two places in *Deep Creek* where Pam Houston shares her shopping lists, including what she planned to prepare for her housemates at dinner. Gay and Houston write of lives surrounded by both nature and people. Though they are sometimes prone to ecstatic reveries, they also deliver quotidian instruction on how to live in the sometimes brutal landscape of our world. In Gay's case: in a Black man's body. In Houston's: as a white woman and survivor of childhood abuse. I love them—both the books and their authors.

"Maybe what I'm missing particularly is the parenting aspect," I told Kate and Suzanne. "Child-free writers versus mothers." The routine tasks that consume a parent. I missed reading about what keeps me from writing one small essay a day.

Kathryn Aalto interviewed me for her 2020 book *Writing Wild*, featuring women who "shape how we see the natural world." It surprised her, Aalto told me, to realize I am the only one of the book's twenty-five women who regularly and openly writes about being a mother.

I'm not judging the reasons a person might not be a parent, or why they might not write about motherhood even if they do have a child. I'm being honest in my own writing about that for which I hunger.

"I think Ellen Meloy and Eva Saulitis both write with quotidian honesty," wrote Suzanne. "But as you say, both were childless. You have me running to my bookshelves!"

When Kate, Suzanne, and I started our text thread in March, patches of never-thawed snow covered my garden, remnants from blizzards that swept over Northern Colorado as the last winter months of 2019 rolled unceasingly into the first freezing months of 2020. Depending on what lies beneath, snow melts at different rates. First to go tends to be whatever snow fell on the hot, dark roofing shingles all around the neighborhood. That drips into our gutters, causing a percussive din for days. Even without the aid of snowplows and shovels, the next to go will be what falls on the heat-trapping asphalt tarring the street. Then our concrete driveways and sidewalks, our flagstone walkway. Snow melts quickly from river rock beds that are still without flowers. Next, any unplanted dirt. Leaving our lawns and flower beds to slowly absorb the remaining patches of snow's hydration. By watching the world around our garden, I've come to understand the ways human population centers create heat islands. A practical climate change experiment in my own backyard.

Before I started gardening, I hadn't paid attention to the sun's influence so carefully. Where it shone longest. Where it never fell.

Before I tried to grow things from the land, the sun was the sun. A constant, distant entity. Powerful and indifferent to my needs. Like God, I suppose, or some people.

In our house's shadow, much of the yard remained frozen that early spring. I hardly ventured into it. Garden people often make it seem as if they spend their whole lives outside the house, poking around flowers. As if they're only truly themselves when breathing the fresh outdoor air. That all other aspects of their lives are unimportant.

That's another limitation.

Like Suzanne, I concentrated on identifying women writers who don't leave their children off the page. Women who also let light shine on that part of their lives. Something in the environmental imagination I was trained in did not admit children, or the women who raise them, into the canon of work that writes about the wild. But a recent emergence of women writes in defiance of such limiting narratives. Barbara Kingsolver's mid-1990s collections are among the oldest. In more than one of her nonfiction books, Kingsolver writes about the family garden and her children—one of whom is named Camille! Jamaica Kincaid's 1999 *My Garden (Book):*

begins with a gift of gardening tools she received on Mother's Day. In her 2013 *Braiding Sweetgrass*, Robin Wall Kimmerer reimagines the ways we might interact with the greater-than-human world. She also writes about lessons she learns as a mother. And, in *World of Wonders: In Praise of Fireflies, Whale Sharks, and Other Astonishments*, published in 2020, Aimee Nezhukumatathil writes about the family she creates with her husband almost as much as she writes about her own childhood and family of origin. Still, what's the old saying? The exceptions prove the rule.

I didn't write anything else to Kate and Suzanne about books that followed or resisted the limitations of this genre, because I got caught up preparing and serving my family's quotidian dinner. Then the quotidian nighttime rituals: combing and braiding Callie's hair; the "did you brush your teeth and wash your face?" cycle; removing still-unfolded laundry from the bed so Callie could sleep. *La la la.* A different kind of woods. The night song of my daughter and me.

Suzanne wrote, "Nothing more natural in humans than the messiness of giving birth. She bustled."

"Drink up, sister!" wrote Kate.

I may or may not have had a drink that night. I didn't write it down. What I know is that in a rush before picking Callie up the next day from what turned out to be her last playdate for many months, I threw a handful of fruit into a smoothie, without turning off the blender's blades.

Next in the text chain: a photo of my cabinets coated in vegetal goo.

Instead of following up on ideas I shared with my friends or cutting back the winter scrabble that wilded our March garden, I focused on the quotidian task of cleaning the kitchen walls and floors and cabinets.

How could I craft a seminal text from a morning like that?

Last season's echinacea pods in spring snow

Ceremony

No one can fly down to bury his aunt.
The sickness is already there. That's what
took her. And, anyway, we are stuck
at home. The moon swelled then emptied
into its shadow. We learned this week
the Black singer died. Days later,
the white one. A man in the neighborhood,
young father of four. Lifted over the sink,
our child stood on the ledge and cleaned
the kitchen windows. It is bright outside
most days. Grass is greening up the yard.
An uncle died. Another aunt was taken
to the hospital. The moon swells again.
This feels like the early days of parenthood.
We swap watch. Focus on raising the child.
We've seen times like this before, we say.
Also, these times are like nothing we have
ever seen. When I came downstairs today
for breakfast, he was playing "Lovely Day,"
a song we danced to at our wedding.
We danced there, in the kitchen, all of us
howling those high and happy days. Lovely
day, we sang. Lovely day. Oh! Lovely day.

The poet's daffodil

Two years after finishing graduate school, I moved to Lynchburg, Virginia, where I stayed from 1999 until 2006. At Randolph-Macon Woman's College (R-MWC), as the school in Lynchburg was called in those days, I taught creative writing, gender studies, and literature.

I enjoyed teaching at a woman's college. Enjoyed living outside the pervasive judgment that so often descends on women who are perceived to take up excessive space in conversations. Studies conducted from the 1960s through the late 1990s revealed that women's participation in conversations could be interpreted as excessive at levels as low as 30 percent. "The talkativeness of women has been gauged in comparison not with men but with silence," said Dr. Dale Spender, a prominent researcher on the subject. "Women have not been judged on the grounds of whether they talk more than men, but of whether they talk more than silent women."

Young women in classes I taught prior to arriving at R-MWC often chose not to share their thoughts much, to avoid the derision that comes with being labeled domineering. But at R-MWC, I regularly heard from everyone in the room, adding to my positive experiences in the classroom. To serve these young women, and feed my own hunger, I built a personal and pedagogical canon of literature that differed from the white male–dominated canon I had encountered at educational institutions up until that point.

I included a poet named Anne Spencer in this new canon.

The Anne Spencer House & Garden Museum stands at 1313 Pierce Street in Lynchburg. Anne Bethel Scales Bannister Spencer and her husband, Edward Alexander Spencer, built the Queen Anne–style building around 1903. For seventy-two years, Anne Spencer lived in the house and worked in her garden. She lived there with Edward while he was alive, with their son, Chauncey Edward, until he married and moved into a house across the street, and with their daughters, Bethel Calloway and Alroy Sarah, until the girls started homes of their own. Anne Spencer died in the house at 1313 Pierce in 1975. Now anyone with an appointment can visit the museum her home became.

I was twenty-six when I first walked through that garden. I saw: the iris; the coral bells; the clematis; the climbing American Beauty roses (lipstick red and multipetaled); the trellis from which spilled Madame Grégoire Staechelin roses (with alluringly scented, ruffle-petaled pink climbing blooms); and the wide-lipped, narrow-throated orange flower Spencer wrote about in a poem she called "Lines to a Nasturtium."

On the double lot between Pierce and Buchanan, Spencer divided her yard into sections she called rooms. In those rooms, she grew a profusion of life from all over the world. Daylilies (from Asia). Hazel varieties native to North America and the United Kingdom. Larkspur, black-eyed Susan, and phlox gathered from the neighboring countryside. She grew hydrangeas (a North American native), snapdragons (an imported southern European plant Thomas Jefferson saw blooming on his Virginia estate around 1767). She planted Virginia red cedar. She cultivated dogwoods (which are native to

North America, not Jerusalem, despite a legend claiming the dogwood flower's four-notched, red-tipped petals are a reminder that soldiers built Jesus's crucifix out of the heart of a dogwood tree). All these different plants from different places foster a sense of radical rootedness on the Spencers' double lot.

I visited the Anne Spencer House & Garden Museum often. Thinking, *Voilà! There is someone who feels as I do.* A woman—a Black woman—lived and wrote and tended a glorious garden in Lynchburg, years before I arrived. I fell in love with the idea of building a home beyond my home while standing inside the garden Anne Spencer built.

Anne Spencer's father, Joel Cephus Bannister, born into slavery a few counties south, claimed Black, white, and Seminole heritage. Spencer's mother, Sarah Louise Scales, was born one year after the end of the Civil War on a plantation just above the North Carolina–Virginia border, where Spencer's maternal grandmother had been enslaved. Anne knew her maternal grandfather was a wealthy white man who claimed a place on the R. J. Reynolds family tree. Her lineage is acknowledged now and was an open secret throughout Anne Spencer's life. But like so many "colored" descendants of America's well-heeled white families, Anne Spencer made a life for herself without the assistance of her grandfather's wealth or opportunities.

Though her body stayed rooted in Lynchburg, Spencer's words remind me of seeds that fly for miles and miles on the wind. Her writing made it to Harlem cultural circles and influential publishing houses on Fifth Avenue in New York. She published poems in the NAACP's journal, *The Crisis*; in Alain Locke's pivotal 1925 anthology, *The New Negro*; and in James Weldon Johnson's 1922 *The Book of American Negro Poetry*. In his introductory notes to Spencer's poems, Johnson wrote, "She lives in Lynchburg, and takes great pride and pleasure in the cultivation of her beautiful garden." She was the first African American woman—the only Virginian woman—to appear in the *Norton Anthology of Modern Poetry*, published in 1973. In 2019, the US Postal Service released a Forever Stamp, part of the "Voices of the Harlem Renaissance" series, featuring Spencer's face. Born from a tobacco plantation's oppression, Spencer chose to oversee growth that brought more life from life.

In one of the garden's rooms, Edward built a small wooden writing cottage for Anne. She called it Edankraal—marrying the first two letters of each of their names, a South African word for "home," and an allusion to the Garden of Eden. In another room, the pool garden, Edward built a pond and fountain with a semicircular stone bench. Overlooking the reflecting pond is a finely shaped black cast-iron bust that Anne called Prince Ebo, gifted to Spencer by her friend, fellow writer and activist W. E. B. Du Bois, who brought the sculpture back from a visit to West Africa. Spencer's garden offered solitude and reflection, but also companionship and community.

In our yard in Colorado, I leave hawthorn and juniper berries on the branch in June for our nonhuman neighbors to consume. When I do this, and when Ray and I make plans to install a bat house in our prairie project, opening the space to nighttime pollinators and mosquito eaters, we welcome beneficial species to help control more destructive ones, following a practice the Spencers believed in as well. Anne and Edward had a pet crow they named Joe, who could speak about twenty-five words. Still, now, an arbor that Edward built drips, in season, with Concord and Niagara grapes Anne allowed only Joe and the birds who visited her garden to eat. The birds entertained her and, she reportedly made sure to point out, they controlled insects. By feeding the birds, Spencer employed what growers call integrated pest management, a system that supports natural food chains rather than rely on man-made poisons to control pest populations. Near the Spencers' two-story back porch, they mounted a martin house on a twenty-foot pole. There, Anne and Edward and the children welcomed migrating purple martins. These seven-inch midnight-blue birds fly up from South America each spring. Along the way, they eat thousands of pesky insects, including horseflies, midges, termites, wasps, and even the flying queens of fire ant colonies. Hosting martins and other beneficial species, the Spencers kept their vibrant yard essentially pest-free.

That the Spencers built such an unthreatened yard wasn't just a matter of aesthetics. Until Edward's family bought the lot after the Civil War, Confederate troops used the land where the Anne Spencer House & Garden Museum now stand as a campground and recruitment site. The Spencer family erected an assembly hall on part of the purchased acreage, offering a safe place for newly freed Black folks to gather. After Edward and Anne built their rambling house and broke ground on the garden, their home remained a stopping place. Their garden, an oasis in an inhospitable land.

During the era of the development of America's motorways and auto clubs and the concurrent culture of long, leisurely road trips, Jim Crow laws and customs excluded Black travelers from the ease of the road. To map out places where it was safe for them to stop, Black travelers relied on word of mouth and resources like the Green Book. Even with these guides, America's highways were often risky and tedious for Black travelers. My mother told me that when her father drove long distances to visit relatives, he asked if his family could use a station's restroom before pumping any gas. Granddad refused to give his money to a business that didn't offer him the dignity of a toilet. In practical terms, this meant he carefully planned travel routes with stations that served Black people in mind. He rarely let the gas gauge drop below half, in case he had to drive farther than planned to find a station where he could safely fill the tank.

As with gas station restrooms, most restaurants and hotels in the United States barred Black travelers. This exclusion was not limited to the American South. Driving through Oregon, Arizona, Illinois, Ohio, Michigan, Maryland, and Pennsylvania, Black travelers packed their own food. They knew in advance which houses or inns allowed them to stay for the night. The Spencer house offered needed sanctuary for touring artists, as well as civil rights activists who traveled from community to community planning strategies with local chapters of organizations like the NAACP. Anne was a founding member of the Lynchburg chapter. Until James Weldon Johnson's untimely death in a 1938 auto accident, the Spencers frequently welcomed the NAACP organizer, writer, and editor, who became Anne's dear friend and literary champion. Langston Hughes slept in the Spencer house. So did W. E. B. Du Bois, Zora Neale Hurston, the Reverend Dr. Martin Luther King Jr., Jackie Robinson, and George Washington Carver. The latter was sometimes called 180 miles north to Washington, DC, to report on his inventions and horticultural discoveries. Highway 29 wound through the east edge of Lynchburg. In the years I drove on that road on my way to DC from North Carolina, I wondered what it was like to be one of those traveling Black intellectuals who stopped to rest at the Spencers'.

Black intellectuals weren't the only ones who passed through Lynchburg. Highway 29, the railway, and the James River brought all kinds of commerce and people. From the late 1800s through the 1940s, Lynchburg thrived as one of the wealthiest cities in America.

The Craddock-Terry Shoe Company, once the fifth-largest shoe company in the world, built its factory in town. Lynchburg boasted an iron foundry, cotton mills, and a paper mill. For four of my seven years there, I lived in a place beside one of the town's many mansions. Back in 1901, the man who owned the trolley companies in Roanoke and Lynchburg built the big house and all its adjacent outbuildings. His workers paved a wide driveway, installing trolley tracks and a roundabout, so his personal trolley car could pick him up each morning and drop him in the evening right at his front door. That's the kind of money and power the Lynchburg captains of industry once commanded.

I rented the bright main and upper levels of the property's old carriage house. A white, retired veteran lived for years in snug basement rooms that must have once been the carriage driver's quarters. My first floor had a sitting room, a dining room, a powder room, and a kitchen. The upstairs was spacious enough for a bedroom, a separate study, and a full bathroom with a clawfoot tub. My bedroom's enormous windows must have been the route through which workers once tossed large supplies of hay. The branches of a big, old deciduous tree waved near my study's window. I kept potted plants inside, including some pothos with vines I let climb up my bedroom's walls and ceilings. Even in winter, I was surrounded by green. For all this, I directed just one-third of my assistant professor salary toward rent.

I can't think of any other Black people who lived along Rivermont Avenue, where many of Lynchburg's grand houses stood, but my job at the college opened doors to me in that neighborhood.

The May I told my grandparents that I'd accepted a tenure-track job at Randolph-Macon Woman's College, Grandma, who nearly always wore a smile, frowned for the rest of the day. Granddad said he couldn't believe they let a Black woman work on that campus as anything but a maid. Before I arrived in Virginia, many Jim Crow laws had been repealed, but forty years weren't enough to fundamentally change a culture built on three and a half centuries of racial segregation and violence. My grandparents had plenty of experience to cause them to be wary. My presence on the faculty was, in fact, almost unprecedented. I was only the second Black tenure-track professor in the institution's history.

For more than one hundred years, Randolph-Macon Woman's College served Virginia's wealthy white daughters, but financial exigency and a shifting mission encouraged the college to foster diversity among the women it served. By 1999, R-MWC actively welcomed young women from places as different from Virginia as Ghana, China, New Jersey, Paraguay, Washington, DC, Staten Island, Jamaica, Kenya, and Los Angeles. Still, only about thirty-five of the school's seven hundred enrolled students identified as nonwhite.

On a warm spring day in 2020, around the time of year when I used to see R-MWC students hang poems from the limbs of a weeping cherry called the Poetry Tree, I texted my friend Vanessa to ask how many of her R-MWC classmates had identified as women of color. Between the two of us, we named most of the African American, Latinx, Asian American, and Native American students who were on campus the same four years as Vanessa. The presence of students like Vanessa—who is Afro-Latina—allowed the college to check multiple boxes without adding multiple bodies to enrollment numbers. We also named several international students, many of whom the college also identified as women of color.

"There were like four international tables in the dining hall," Vanessa texted. "It may have been three."

The eight-seat tables in the campus dining hall boasted classy centerpieces with round vases and seasonal flowers. Chrysanthemums in fall. Daisies in springtime. A predominately Black dining staff cooked for the students and set up these attractive tables. African Americans made up 30 percent of Lynchburg's population when I lived there, but hundreds of years of subjugation and violence barred many of the city's Black citizens from accessing the tools that would have helped them to sit at those tables.

Still, some Black citizens found ways to work around the systems established to keep them from thriving. Edward Spencer, Anne's husband, was the first African American postman in Lynchburg. A job that offered financial security and useful connections. Lining his route were big houses like the trolley owner's mansion I lived beside. When the wealthy white residents of these homes threw some still-valuable item away, Edward hauled the castoffs to his family's place on Pierce Street. A twelve-foot-tall gold leaf mirror from one mansion's ballroom. A stained glass window from an entryway. Some Gothic wire screens. When Randolph-Macon Woman's College built a tall redbrick wall around the campus, Edward acquired part of the college's original lacework cast-iron gate, installing it, like other refurbished pieces, in the Spencers' house and garden. Their house was dazzling outside and in. Not because of flashiness, though Anne and Edward did have bright flare, but because the house was so clearly the site of deep care, bold ingenuity, and renewing love.

The need to make a bold, fresh way through seemingly pathless terrain is familiar to many Black women. Denied access to formal education by the segregated systems of the time and region, Anne was barely literate when her mother sent her to Lynchburg in 1893 to attend a school known at the time as Virginia Seminary and now known as Virginia University of Lynchburg. The institution served Black youths from grade school through and including college. Once enrolled, Anne quickly learned to read English as well as French, Latin, and Greek, graduating as the valedictorian of the 1899 class. During her time at Virginia Seminary, she met Edward, a fellow student. Anne helped him with classical languages. He tutored her in geometry. But when white patrons visited the school, students put books away and focused on the trades: sewing, carpentry, and bricklaying. These patrons expected Virginia Seminary's Black graduates to be humble, not high minded.

The way some biographers tell it, Anne Spencer created her intellectual life, and her garden, almost entirely outside custom and tradition. Spencer's garden designs may have been influenced by popular, white-focused home and garden magazines of her time, but she made them her own. Her imagination triggered what she built. In her neighborhood and in Seminary Hill, the professional and working-class Black neighborhood near her former school, Black gardeners grew food to supplement their families' tables. In her own garden, though, Spencer grew only a few herbs for human consumption. She focused, instead, on how beauty fed her.

But flower beds surrounded many houses in Spencer's neighborhood and up on Seminary Hill—with hollyhocks, bright orange and yellow nasturtiums, azaleas, and sweetly scented white and pink and lavender peonies. Black people organized garden clubs. Black people hosted lawn parties. I believe a large part of the cultivation of the idea of Spencer's singular gardening achievements is because her white biographers did not know about, or pay attention to, the Black gardeners who lived alongside Spencer.

My own grandparents taught at Virginia Seminary in the 1940s and '50s. Their experiences in Lynchburg are why they responded so strongly when I told them I planned to move there myself. They had lived three blocks from Anne and Edward Spencer's Pierce Street house. Granddad pastored an American Baptist congregation with a historic brick church just a twenty-minute walk from their home. This meant Granddad didn't have to ride the segregated trolley car to and from the Court Street Baptist Church building.

Back when my grandparents lived in Lynchburg, Black people weren't allowed on the R-MWC campus except to provide menial labor in the buildings or on the grounds. Barred from the town library, Granddad once asked permission to check out books from the college's collection. The librarian said she could retrieve certain volumes, which she would allow him to read in the basement furnace room. Granddad refused. Until the day I told my grandparents I'd accepted the job in Lynchburg, I'd never heard this. They thought they moved away from that city and its ugly stories once and for all. Why would I want to live there?

The work required to diversify a landscape can take so many people—so many years—that by the time the desired changes develop, they can be hard to recognize or believe.

My mother and aunt were born in Lynchburg and completed most of their elementary education there. The family didn't stay in Lynchburg long enough for Mom to attend Dunbar High School, the Black school where Spencer served as a librarian. Still, Mom remembers hearing Spencer's name. Anne moved in the same social and cultural circles as many of the adults my mom loved. The adults who fought to protect my mother's life and their own.

I need to say this directly, to honor actions the phrase "fought to protect" might elide.

As in many American towns, above and below the Mason-Dixon Line, racial segregation in Lynchburg meant white ambulance drivers, as well as guards, nurses, and doctors at the hospital, refused Black patients. If they deigned to treat Black patients, they might do so in a spare room near the morgue. They might use unsanitary tools. They were likely to offer subpar attention. My father tells a story from his medical training in 1960s Chicago. Over the top of the doorframes in delivery rooms at a Black-serving hospital, white obstetricians posted plaques that read, in Latin, ILLEGITIMI NON CARBORUNDUM: "Don't let the bastards grind you down." A joke that reveals deep distain and disregard for Black parents and their children. By treating patients in such an inferior and contemptuous manner, those who worked in the hospital system often killed Black people. That's the inevitable and often intentional outcome of racial segregation and bias.

Composed of individual men and women—like Anne Spencer and like my mother's godmother, Alice Gilliam—organizations like the NAACP fought to protect Black lives. These men and women and children wrote letters and marched and set up phone trees to efficiently share information and bailed protestors out of jail and walked their own bodies through spitting mobs and formed picket lines between segregated businesses and slur-casting crowds and mimeographed thousands of flyers and made meals for busy organizers and made more meals for grieving families and provided their cars as alternative transportation for people boycotting racially segregated transportation lines and opened their homes to Black travelers and sent flowers cut from their own gardens to meetings and funerals because the white florists would not sell flowers to Black people so Black people grew their own beauty and dug in and continued digging wherever and however digging was needed.

Not willing to risk the hospital in Lynchburg, my grandmother delivered my mother in Alice Gilliam's house on Seminary Hill. A house surrounded by redbud trees, peonies, hollyhocks, forsythia, and nasturtium.

Walking in a municipal garden here in Fort Collins one late May afternoon, I pointed to a bushy burst of new green stalks. Without hesitation and with a huge smile, Mom said, "Oh, those are peonies." They remind Mom of her godmother, Alice. She loves the forceful, steady perfume of a peony flower to this day.

I keep a photograph in my hallway, taken in the sitting room of my grandparents' Lynchburg house around 1952. In a few months' time, multicolored peony bushes would bloom in Godmother Alice's yard. But my mother's family took the photograph on Easter morning, when the world around them still held on to the cold. In town: white-only drinking fountains and colored drinking fountains. White-only swimming pools. White-only diners. Stores refused to allow Mom's family to try on clothes. The houses of God were segregated too.

In the late 1840s, a century before the photo I keep in my hallway was taken, Black men and women established the church my grandfather pastored. Then, between 1879 and 1880, Black laborers built the formidable brick building on Court Street. They constructed their own house of worship. A sanctuary apart from the white men and women who built a system of wealth and power dependent on Black enslavement and debasement and toil.

All over America, Black people constructed such safe spaces by and for themselves. By order of the deed of the building that housed Lynchburg's public library, my grandfather wasn't allowed to visit the library's premises or use its collection. So the members of Court Street Baptist Church started a library fund to help my Black mother's Black father buy the books he needed to preach the sermons his Black parishioners needed to hear every Sunday in that deeply segregated, often hateful town.

I love this picture of my mother with her family on an Easter morning in the 1950s Jim Crow South, just miles up the road from Danville, Virginia, the last capital of the Confederacy. My dear grandmother, both her girls, in their Easter hats and bright dresses. Carefully draped sheers on the window behind them protecting the interior of their house. Grandma in lace gloves, with a pearlescent necklace and matching earrings, perched daintily on the arm of her husband's lounge chair, hand touching his. Granddad, in the chair, legs pressed together, hair shorn close to his scalp. He wears a black suit and clerical collar.

My aunt Jeannye is two years younger than my mother. Her crossed legs and the casual tilt of her right foot suggest my aunt's shoes were often unevenly scuffed. Her right shoulder is endearingly slouched, in a position that looks more comfortable than poised. She has the bearing of a girl still allowed to be a child. During a phone call one pandemic summer Sunday, I mentioned the photo and her stance.

Aunt Jeannye said, "Did you know I'm really shy and very private?"

In all the years I had this picture on our wall, my aunt being shy never occurred to me. Why didn't I notice the way the little girl pulls from the outside gaze? It must have been distressing to always be scrutinized so closely, often by hostile eyes.

But my mother, no older than eleven in the photo, looks like my mother. Poised and composed and diligent. Unflappable.

The world weighs on each of us, but in unequal measure.

Someone pressed and neatly braided my mother's shoulder-length hair. With her hair pulled back, I can see my grandfather's features reflected in my mother's face. I love how my mother stands close to her own mother, as she continued to do for the rest of my grandmother's life. In the photograph, Mom is set a little bit behind her family. Not like a shrinking violet. Instead, with confident security in her stance. I bet she had already helped her mother accomplish some monumental administrative task that morning—mimeographing church programs or folding bulletins. Or maybe she'd helped get breakfast on the table, so her mother could type the last page of my grandfather's sermon. Even as a child, I think my mother must have been called on to complete difficult tasks with a smile.

In her smile, in this photo, I can see my mother's connection with the people who love her and whom she loves. I can see the love I was born from and raised into. When I walk past the red, yellow, and purple flowers on the tulip- and vinca-lined path to her brick house, my mother opens the door to me and gives me that same smile. I don't know who held the camera that day, but her smile is for that person as well. Her smile has been the same for three-quarters of a century. Look at the young girl's plump cheeks. She's still got those too.

By the time I walked in the Anne Spencer Garden, twenty-five years after her death, Anne's son, Chauncey, took on much of the work required to maintain his mother's legacy. He led the museum's board and organized the garden clubs and contractors who refurbished the house and the yard. While Spencer lived, she hardly let anyone but Edward work in her garden. Fearing they'd unwittingly kill something she loved, Spencer allowed her children to pull only dandelions and other highly recognizable weeds.

I can understand Anne not wanting others' hands in her garden, but at some point, other people's spades must turn up the soil where a gardener plants her dreams.

Gardens are always, by their nature, wholly original. The mind spring of the gardener who puts in the work. But gardens are also dependent on the labor, the knowledge, the propagation of those who come before, alongside, and after.

During Anne Spencer's final illness and throughout the eight years following her 1975 death at ninety-three, her garden fell into decline. Invasive honeysuckle and poison ivy took over. Rosebushes thickened into self-choking masses and hardly bloomed. The fountain stopped circulating water and Prince Ebo stared into a stagnant pool. The pergola buckled under the weight of never-thinned wisteria vines. The neighborhood around the Spencer house buckled as well. Iron foundries and tobacco warehouses and the leather shoe company and trolley car line all collapsed. Taking with them jobs that supported local families. Dr. Walter Johnson, who coached the tennis stars Arthur Ashe and Althea Gibson, died. Other Black professionals moved away. But, in 1983, a restoration project began on the Spencers' double lot. Because gardens are living, growing, demanding, and dynamic, the restoration continues.

Gardens, history, and hope are the same. Though once dearly beloved, if left untended, without anyone's dedication and care, much will be totally lost.

In 2015, a women's college located twelve miles up the road from Lynchburg declared bankruptcy. As Sweet Briar College prepared for dissolution and the property's sale, the college's predominantly wealthy white alumnae voiced their distress. Another group of people held deep concerns about the future of Sweet Briar as well: the descendants of 115 people once enslaved on the three-thousand-acre Sweet Briar Plantation.

During Reconstruction, one of those formerly enslaved men, James Fletcher, bought some neighboring property. The cabin he built still stands, and many of his descendants live and work in the immediate area. Other formerly enslaved men and women, and their children, moved away. To Washington, DC, and farther north. To the West, perhaps as far as Colorado, California. Despite this diaspora, "amongst the hourly wage-earners at the college," a 2015 *Washington Post* article reported, "nearly a third are directly descended from Sweet Briar slaves." More than some college girls who spent four years at Sweet Briar, these Black men and women trace generations to that stretch of land. Descendants from all over the country visit a burial ground on the property during annual family reunions. They visit their enslaved ancestors' remaining cabins, now historic sites. They return to walk on land the Fletcher family and others tended for generations.

Every time somebody, Black or white, says that the attachment I feel to land is strange, I think of those Sweet Briar Plantation families. Their fidelity to that place. Fidelity born from lack of opportunity and enforced by segregation, systemic racism, and often brutality. But that is not all. There is also a deep, abiding joy—an emotion derived from a sense of comfort, success, and accomplishment. On that land, men and women raised the next generations with an eye toward, and then in a state of, freedom. They fought hard to claim their welcome.

I've seen dogwoods bloom in the Virginia springtime. Their

nearly excessive displays. Those notched crosses of white and pink petals. I've seen redbuds shoot out unrepentant bright buds, like a calendar, around Easter. Azaleas bursting out in technicolor purple and gold and magenta flower. Forsythia mimicking the sun, golden and dazzling. The plants in Virginia alone are enough to encourage a seeing person's site fidelity. Those plants grow up from land that generations of Black people carefully cultivated and tended and loved.

My grandfather, the one who lived for a spell in Lynchburg, grew up just outside Baton Rouge, Louisiana—in a community known as Scotlandville. Driving long hours on carefully plotted routes, he took his family back there most summers.

My mother remembers a grove of pecan trees in the yard there. She remembers a napping porch where they slept on hot summer nights. The DDT trucks whose cooling spray children ran behind on the adjacent road. Each night before going to bed, her grandparents sprayed their entire house to get rid of mosquitoes. Poison swirled all around them. Mom says, "It's a wonder we survived!"

My grandfather's father was a farm demonstration agent in the Baton Rouge area. My great-grandmother taught elementary school. The two met at nearby Southern University, a school founded as Southern Agricultural and Mechanical College in 1890, as part of a second Morrill Act. Unlike the first Morrill Land Grant College Act of 1862, the 1890 act required states to build institutions for Black students. These and a few schools founded by religious denominations or wealthy white patrons, such as Virginia Seminary, were the only higher education institutions most Black students could attend. Both of my grandfather's parents graduated from Southern.

So did my grandfather and his two brothers. Continuing a tradition set by his parents, Granddad met my grandmother while the two were both students. My grandmother's aunt Mary Carlisle Meadors served as registrar for the college from 1920 to 1953, and so Grandma and, over time, five of her twelve siblings, traveled to the college from a different part of the state. My family believes that education changes lives for the better.

As a state Farm Bureau agent, my great-grandfather drove country roads in southeast Louisiana. He taught young Black men the best ways to plant crops. He sometimes shared his knowledge with rural white men too. Mom remembered learning early that this wasn't always easy for her grandfather. Once, refusing to listen to his lessons about crop rotation and allowing fields to sometimes lie fallow, a gang of white men encircled my great-grandfather's new Ford. They wouldn't be taught how to prosper by an "uppity" Black man. The white men told my great-grandfather, "A horse and cart will do just fine next time you come around these parts."

My great-grandfather died before I was born.

What can I say I learned from him?

The pecan grove in Scotlandville has been gone a long time.

And what about Granddad's brother, my great-uncle Hugh? He left Scotlandville and traveled annually to Fort Collins in the early 1950s. What was a Black man from Louisiana doing in Fort Collins way back then?

Soon after Ray and Callie and I moved to Fort Collins, the Pioneer Association commissioned a plaque for a Cherry Street cottage—a childhood home of Hattie McDaniel, the first Black person to win an Academy Award. As early as 1900, a few Black families clustered in a block of cottages a mile from the grand homes on Fort Collins's oak-canopied West Mountain Avenue. McDaniel's family lived among them. A photo in the *Coloradoan* newspaper's archive shows me holding Callie on my hip while I read the new marker. Callie's dress and parka sport bright floral patterns. Though it is not a story America often tells, McDaniel's time in Fort Collins reminds me that Black people are, and always have been, planted everywhere in this country.

For almost seventy-five years, the 1896 US Supreme Court decision known as *Plessy v. Ferguson* allowed states and institutions to maintain racially separate facilities. The law promoted the segregation of busses, trains, trolleys, restrooms, libraries, schools, hospitals, clubs, concert halls, museums, swimming pools, movie theaters, housing, workplaces. Everything. If they funded schools for Black students at all, white administrators and legislatures underfunded them. Anne Spencer served as the librarian at Lynchburg's Dunbar High—a school named for the revered Black poet Paul Laurence Dunbar—but when she started her job, she arrived to find no books. To create a library, Spencer donated her own collection of books. If she hadn't, most of Lynchburg's Black community would not have had access to any. Black people who insisted on thriving exercised this kind of resilience and ingenuity over and over. Because all the graduate schools in Louisiana and its surrounding states barred his admission, my great-uncle Hugh traveled to Colorado A&M, known now as Colorado State University, to pursue his graduate education in agriculture. A separate-but-equal loophole funneled hundreds of motivated law, MD, master's, seminary, and PhD candidates—students like my grandparents and nearly all their siblings, including Great-Uncle Hugh—to universities in places like Iowa, Nebraska, New York, Wyoming, Colorado, and even England and Europe.

Maybe my people are like dandelions, planting ourselves where the earth offers openings. Or maybe we're more like irises—able to withstand division. Thriving and flourishing in many climates and soils.

"No," Ray said one mid-June 2020 evening. "We're like dandelions. When they see us, they still try to kill us."

My parents tried to find a record of Great-Uncle Hugh at the CSU alumni center. He graduated in 1953, completing all his coursework during the summers, when he was not teaching his own students back in Louisiana. Mom and Dad hoped to locate yearbook photos or archival documents that showed some evidence of his presence on campus during those years, but when I asked them to tell me about their search Dad said, "What we've discovered is that these people who attended in the summers are virtually invisible."

In a 2018 *Oxford American* article about Anne Spencer, I praised the fact that Spencer "was working at the intersection between environmental and civil justice." Her work—as a librarian, a civil rights activist, a mother, a wife, a writer, a host, and a champion for the human and nonhuman lives around her—seems always and fundamentally connected.

Spencer was able to integrate the many threads of her life, in part, because of her garden. In the book *Half My World*, a title taken from a line of Spencer's poetry, Rebecca T. Frischkorn and Reuben M. Rainey write, "She used her garden as a restorative sanctuary to which she could retreat in fatigue, frustration, or sorrow, as well as in celebration." The book's dozen photographs of Spencer in her house or garden all show her smiling. The kind of smile my own grandmother frequently wore. The kind my mother wears still. The kind that floods the body with joy's relief. The photographs span decades, but always that smile. Solid and monumental, resisting the dissolution of degradation and despair. In her home and garden, Spencer built a place that made her smile.

As with the record of Great-Uncle Hugh at Colorado A&M, of the pecan grove on my great-grandparents' land, and of the individual stories of the 115 enslaved people who lived and worked on the Sweet Briar land, records of Spencer's life are scarce. During her long final illness, intending to help clear out what they considered clutter, people discarded a lifetime of Spencer's papers. Those of us hungry to learn more from her life must make do with the scraps that remain. And what remains of her garden.

I stayed in Lynchburg only seven years, after which I moved back to my own home places—first California, then this house in Colorado. But there is not a week that goes by when I don't think about that town in Central Virginia where Anne Spencer built a life, wrote her poems, and built a garden. Where my grandparents spent ten years building lives for our family too. Not a week where I don't think about what it means to grow something beautiful from what might seem like dirt.

I have a friend back in Lynchburg who worked in Spencer's garden after it became a museum. Raised by Mennonites, my friend has a way of sitting silently in a garden that helps me understand how I might open myself up to hearing the welcoming wonder of this world. Sometimes I think to call her and ask for clippings from Spencer's garden, something I could plant here in Colorado. But I have enough of those memories rooted, already, deep inside me.

Snapdragons grow in our yard and in Anne Spencer's garden

In the fall of 2019, before we started on the prairie project, our neighbor Pam brought me hollyhock seeds she'd collected on a walk. Dark, round, and bumpy, like dry chickpeas, the seeds need light to germinate. I planted them only a quarter-inch deep.

"They get really big. And they come back year after year," Pam warned. "Be careful to put them only where you want them."

I planted one along our backyard's east fence, around a cluster of early-blooming bearded iris, and one in the reclaimed bed near the northeast corner. Both those hollyhocks stay short, no more than three feet tall. They produce modest flowers. Striking up close. Breathtaking, really. The deep purple of claret in a glass. But, from far away, almost invisible.

The third seed went into a patch I began to call the dooryard once the hollyhock bloomed. First to emerge were low, wide leaves the size of salad plates. Then the stalks grew taller than me, to nearly six feet. Each stalk boasted bright green acorn-size buds all along thick, sage-green stems.

Eventually, several palm-broad, creamy yellow and girl's ribbon–pink flowers opened up and down each towering stalk. Like full skirts, the petals flared around bright yellow stamens and pistils—the strawlike bits that constitute the flowers' reproductive parts.

The flowers' centers darkened. Bull's-eyes for bees. So sensual, these hollyhocks, and showy. My God.

One of those stalks tended to wander, reaching like a tendril from Jack's giant beanstalk toward our front door. If you rang our bell on a windy day, that hollyhock might smack you in the face. I have to train it, say the garden books. Tie the stalk to keep it in place. Unruly plants betray an unpracticed gardener.

I don't want you hurt, you who visit my garden. I want to offer a welcoming space.

I'd come to the end of the time I could dedicate to writing one afternoon, when this paragraph tumbled from my fingers:

> None of these are accidents: the omission of Black and Brown stories from literature taught in schools set up to serve white people; the condescension my professors conveyed when they considered a Black woman's writing; the absence of stories in canonical environmental prose of women actively engaged in the work of mothering; the prioritization of narratives of solitary men in the wilderness. These conventions are part of a design.

I wanted to stay at the desk, to dig more deeply into what the words meant. But I had to collect Callie from school. This was in February 2020, before the disruptions of remote schooling and business closures and rampant wildfires and the smoke and the dead and the dying that changed the ways and truths and timelines we taught children.

So often, I put aside an idea then pick it up again months later, rearranging the order of telling to more accurately reflect the warped arc of time as I live it. That afternoon, in early 2020, I washed the dishes and prepped dinner while Callie focused on her homework. Then I ferried her to the dance studio. While she danced, I scooted over to the gym to work out before driving us home, serving the family the dinner I'd begun earlier, reading with Callie for a half hour, and then seeing her off to bed. I did not likely stay awake much longer. If I got back to the page at all (in this case, for many months, I didn't get back to that page) it wouldn't have been until after the morning ritual, and then only if I didn't have to go to campus to attend to what might be called my "actual job." My job as a professor that, among other things, helps us pay the mortgage and maintain health insurance.

For the genre-shaping writer John Muir, these were not concerns. Muir—bless his trailblazing, Sierra-loving heart—rejected the agrarian life of his father, married a wealthy concert-level pianist who refrained from playing piano when he was at home so as not to disturb his writing, and kept a Chinese cook, who made sure Muir had food when he chose not to live on stale bread for days as he tromped through the mountains taking notes for his books.

The social and financial position afforded by Muir's writing and his marriage meant he could speak directly to the ears and pocketbooks of the captains of industry whose clout and cash make things happen in America. Even on a 1903–4 world tour, during which Muir encountered stands of "mints, poppies, hollyhocks, etc., in glowing profusion" while walking in "the only mountains seen since leaving America," when John Muir got lost in nature, he could find himself again. As self-possessed as ever. Then he could wander home to his California estate overlooking a magnificent garden and orchard. Muir profited from those fruit and nut orchards, though for many years he relied on relatives and the men his relatives hired to do the tedious work of tending to the trees for him.

I have a pecan I gathered in that orchard years ago. It rests, maybe rots, on a bookshelf in my study. I have a bookshelf in a study in a home I own. I have funds to order Chinese food if I don't feel like cooking. I have a spouse who works to make a quiet space in which I can write. In such ways, my experiences are not different from Muir's. In other ways, mine is a life he never imagined.

By the time I got back to the page I started in February 2020, it was September. I loaded and unloaded the washer and dryer in breaks between videoconferences and checking on Callie's progress on her remote school assignments. At dinner, Callie panicked because she had forgotten about her book report. I stayed up with her until eleven p.m., and she still needed to wake up early if she had any hope of completing and submitting the project on time. Meanwhile, a box of tulip and daffodil bulbs sat on my gardening table with a sheet of instructions urging me to get them in the ground immediately. Oval orbs with roots dangling from the bottom like tiny yellow onions. If given enough time to acclimate to new beds before long winter months prompt their dormancy, these bulbs will grow green stalks and sprout flowers by spring. In the early hours of that bright fall morning, I could sleep or write or garden or clean house or help my daughter, but I couldn't do all or even most of those things.

Which needs, which lives, which wills, which words, which stories, ought to have priority? These are questions I return to, like I return to these pages about my garden, puzzling over what gets my attention, and in which order. I typed as quickly as I could, revising my focus away from the frustrations I penned down in February.

I tried instead to describe the hollyhock that finally offered the blooms I'd hoped for all summer.

Until September, there were no flowers on the hollyhock in the dooryard. I kept meaning to find out what was wrong, but whenever I had a moment, life demanded something else. The morning I wrote rather than planting bulbs, Callie woke up, still worried about her book report. A fifth grader, she had to do the work herself, but someone needed to check on her progress.

Again, I stopped typing.

A 2008 revision of the *Oxford Junior Dictionary*, a reference book aimed at children ages seven to nine, removed fifty words, including *minnow, mussel, ferret, wren, bloom, sycamore, heron, beaver, lark, magpie, bluebell, buttercup, carnation, clover, crocus, dandelion, lavender, pansy, tulip, violet,* and *blackberry*. So much earth-grown language.

Language helps us shape our imagination, helps us shape perspective. Language helps us reach toward empathy and understanding. Language helps us learn about history, our place in the now, and possibilities for our future. A future where there may, in fact, be fewer sycamores, fewer varieties of buttercup, fewer mussels, fewer herons, fewer of what pollinates and thrives on clover, fewer beavers, fewer magpies, fewer blooms. Or, perhaps, none at all. How damaging these omissions could be for the development of the scope of children's imaginations and their connections to the living planet.

The lexicographers' defense made a kind of sense. This dictionary contained only about ten thousand words. Language changes. The words children need to navigate the world also change. New words demanded attention, the lexicographers explained. *Blog, broadband, celebrity, compulsory, vandalism, voicemail, chatroom, biodegradable, endangered, cautionary tale*, and the already essentially extinct capital B *BlackBerry*. The words that replaced earth-grown language connected children to commerce, urban living, the human-centered indoor world where many of us spend so much time. Farewell to the days of blooms and bluebells.

If I limit my language, the way I articulate my perceptions of the world, certain ways of being become impossible. If I freeze the dooryard hollyhocks in the moment before they grew unruly, I erase the vision of their whipping frenzy in the wind. The way they fling their seeds far from their pods. The way they dance along with the rhythms of the breeze.

Upheaval is a word rooted in the soil. It suggests the act of turning up the earth to make room for sowing. After upheaval, new life is revealed. Earthworms. Millipedes. The sow bug's name isn't so much about the piglike look of the insect as it is about the fact that the upturning of earth that comes with sowing and planting tends to bring sow bugs to light.

Here is something I find interesting: I never saw myself as an animal so much as when I hunched on all fours to labor with my girl. Or when I nursed her, my daughter's newborn suckling frenzy more feral than anything I ever held so close. I was a mammal mother. A female beast. Since then, I have not fully returned to the independent person I believed I was before.

A family of blue jays nests in a yard behind our house and wing through our airspace most days, flashing their brilliant feathers. In the pandemic-shaded September of 2020, there were also many hundred thousand acres of wildfire burning close by. On one of the smokiest days, when the skies looked like *The Day After* we practiced for in the nuclear-threatened 1980s of my childhood, the blue jays flashed across our yard looking quite plain. I had to point the birds out to Callie, who did not recognize them. Untrue to the name by which we've come to know them, the birds we saw looked only white and gray. The jays' feathers carry no blue pigment. Our eyes trick us into seeing that shade when light reflects off the birds. On that red-orange day, when the light was so altered, we had to train ourselves to see the birds anew. Maybe this is what it means for me—for this Black woman, this mother, the subdivision dweller with a little yard she tends—to say I write about the wide wild world. Maybe it is just a matter of figuring out what can be seen in this light.

The word *eco* derives from the Greek *oikos*, meaning "house." In the context of the word *ecology*, we understand that house to mean the space of nonhuman living entities and their environments. In turn, *environment* suggests a *natural* world that is larger than and separate from the human. The world of bluebells and beavers and buttercups, not broadband and blogs and the capital-B BlackBerry. The *environment* is *outside* the door.

In nineteenth- and twentieth-century environmental literature, the environment is often synonymous with the wild. "A state of nature not tamed or domesticated," according to the dictionary on my desk, "an uncultivated uninhabited region." A space far, far outside the door.

"It is important not to confuse the wilderness image with the garden or pastoral image," writes Leonard Lutwack in what one review calls his "groundbreaking study," *The Role of Place in Literature*. "The essential difference is that land forms, plants, and animals remain in a natural condition in the wilderness," Lutwack continues. "Trees, rocks, and wild animals are proper symbols of the wilderness; fruit, trees, flowers, and domestic animals, of the garden." The division Lutwack describes assigns some "natural condition" kept at a remove from the spaces humans inhabit. But I want what is inside my doors to be part of this conversation. I don't want to separate my life from other lives on the planet.

Ecological thought, conservationist thought, the thoughts of the gardener—these should foster nurturing and collaborative relationships with other life-forms, including those we've long called wild. This planet is home to us all. All who live in this house are family. What folly to separate the urgent life will of the hollyhock outside my door from the other lives, the family, I hold dear. My life demands a radically domestic ecological thought.

My childhood home backed onto a hillside covered in chaparral. We call it *chaparral* because sage shrubs and manzanita bushes and purple needlegrass caught on the *chaps* of Spanish conquistadors as they rode through California's hills during those western waves of colonization. I was a Conquistador all through junior high. That's what we called ourselves. The Rancho San Joaquin Conquistadors. Without awareness or apology. I ran through the hills behind my house most afternoons. Burrs entangling my hair, my clothes, my memories.

In the Bible I read during those years, *wilderness* refers to the world *outside* the Garden of Eden. A place of trials and tribulations. Of worms and deadly beasts. A parched land to wander in fear, where only a miracle could coax a crocus from the ground. One origin of the word *wilderness* comes from the Old English *wilddēornes*. The place of the wild deer (*dēor*), where untamed animals roam. I walk through a world where to say *nature* is to mean this kind of wilderness: "an area where the earth and its community of life are untrammeled by man, where man himself is a visitor who does not remain" (the Wilderness Act of 1964). A world of ideation caught up in Biblical burrs.

But the word *wilderness* once also applied to the garden. In seventeenth- and eighteenth-century gardens, plants that didn't fit in the ordered alleys and beds of the formal estate landed in the wilderness. Unruly rosebushes and wild-seeming shrubs. Statues of the wild Bacchus, and of Pan, that wine-swilling half-goat, half-man. My computer's dictionary defines *wilderness* as "an uncultivated, uninhabited, and inhospitable region," but also "a neglected or abandoned area of a garden or town." A wilderness can exist in our own backyards.

The parcel of Southern Californian land where I grew up was semiarid desert for twelve thousand years, until the Irvine family took it as their ranch. The University of California system built a campus on a portion of that ranch. Soon after, my father left his faculty position at the University of Colorado's Denver-based medical school to join UCI's faculty. In the early 1970s, just over thirty thousand people lived in that near-coastal California town. The ranch converted many of the lowland valleys into strawberry and asparagus fields, importing shaggy eucalyptus to plant as wind blocks along the edges of irrigated rows. The air often smelled of steamed green asparagus. Or the crisp menthol perfume of eucalyptus. Or the metallic taint of the biocides sprayed over the fields. But native sage and manzanita still covered the hills that backed our house, scenting our yard with the dusty, sweet spice of their bark and the honey-thick smell of small flowers. Bright orange California poppies burst open after winter rains. On breezy days, I'd hear the rustle of the Christmas holly–shaped leaves of desert live oak.

Riding in the car along Red Hill Avenue as a child, I saw the summer hills wearing dry russet grass like cashmere sweaters. Quail scurried in and out of chaparral near what's now the Quail Hill subdivision. To build roads and houses and commercial structures for the growing population, bulldozers leveled wide swaths of the hillsides' sage. By the mid-1980s, when we left, about 100,000 people lived in Irvine. There was less chaparral to stabilize the hillsides and prevent soil from sliding into the valleys after hard rains, so developers planted quick-spreading ice plant, imported from South Africa, on newly terraced slopes. An effort to restrain disrupted ground. In 2020, about 290,000 people occupied the same sixty-six square miles where I lived as a child. When I visit, I barely recognize the wild hills I once considered home.

Fort Collins reminds me of Irvine in the 1980s. Mostly, I mean this as a compliment. I love the open spaces designed into the town's development plans, the evidence of children in our many parks and high-quality public schools. I feel comfortable in this suburban peace. I enjoy our shopping districts and two-hundred-plus miles of bike trails. But when I say my new home reminds me of my childhood town, I understand the implications of so many people coming to love the same place. I mourn for the bulldozed and Caterpillared hills of my childhood.

When we moved here, 150,000 people lived in Fort Collins. Four times the number who lived here when my great-uncle Hugh finished his graduate degree. More than double the number who lived here when Mom's best friend, my aunt Mary, moved to Fort Collins in 1975. Less than a decade since our 2013 arrival, the population has grown by an additional 15 percent. On bike rides home from work some warm September evenings, I pass an open field covered in the bright green leaves and small purple flowers of wild alfalfa (*Medicago sativa*). Used sometimes as forage for livestock, and often just growing wild, this drought-tolerant species was introduced from Eurasia three hundred years ago and has made itself at home here in the Mountain West. The fallow field's alfalfa flowers welcome so many butterflies and bees that I can see their wings flap from plant to plant even as I whiz by on my bike. But a bright yellow NOTICE OF INTENT TO DEVELOP stands in the field, and I know someday I'll bike by to find this bee-glad glade bulldozed and built over.

By the time Callie finishes college, town leaders anticipate that 300,000 people will live in Fort Collins. I know this kind of erasure. Acre after acre, I've watched open land be built up. With all these new people and new human buildings and turfed parks and paved roads, there will be less space for black-and-white magpie, red-winged blackbird, prairie dog towns, and bloom. I love to see these thriving in the open space along the trails around my neighborhood. But unless my community and I make active and careful decisions about what kind of life we support in the place we call home, I can't count on them to be here much longer.

Fort Collins exists in a part of Colorado I refer to as *Nebraska . . . Nebraska . . . Nebraska . . . MOUNTAINS!!!* Driving in from the east over the plains—through cattle feedlots, fracking rigs, and fields— the air smells of sweet, sharp methane fumes and grassy, pungent manure. But just three miles west of our house, red-rock outcroppings that folks here call Devil's Backbone jut from the earth, abruptly separating the grassland, feedlots, and prairies of western Nebraska and eastern Colorado from the part of Colorado made up of fourteen-thousand-foot peaks. Riding a bike into those foothills and the Front Range of the Rocky Mountains requires the thigh muscles of a certified athlete. Wind that blows in from the west smells like juniper Christmas candles and herby, peppery sage.

It took me several years to identify the source of those smells. I am still learning the names of much of the flora and fauna in this part of the world. During my first years back in Colorado, this agitated me. I knew how to *see* the place I called home, but I needed to learn so much more if I wanted to build the kind of life I wanted here. In part, this book is my family's instruction manual. In part, it's our deed.

In May 2020, I joined several groups on social media dedicated to Black gardeners. I wanted fresh interactions during my isolated days. I wanted to know what grew in climates where gardens had already come to life. In the gardening groups' safe online spaces—designed for Black gardeners because so few spaces in this country are safe for those who identify as Black—people ask questions, discuss setbacks, and brag about growth.

Visiting the groups, I found myself surprised by how frequently posters led with fear. People who hadn't gardened before expressed fear about killing their plants, fear of choosing the wrong plants or not planting them correctly, fear of interacting with the insects and snakes and frogs that shared their plots. So much articulated anxiety. It took my breath away. How had so many people been conditioned to doubt their ability to achieve success in a garden?

"I'm nervous as eff," wrote one local friend in a private message, "but this is something I've been wanting to do for a while now." She planned to plant tomatoes, basil, and zucchini. Marigolds and nasturtium.

"What's the worst that can happen?" I replied. "Everything dies and you have to go to the grocery store? That's not really so awful. But if things grow! Delicious and nutritious!"

I wanted to encourage my friend's excitement. I wanted her to trust gardening as a source of extreme pleasure. I wanted her to understand that learning on this job came with little risk and the possibility of great reward.

When I first got my hands on this garden, I didn't know much about what did and did not grow well in Northern Colorado's conditions. In the summer, the sun shines viciously. But as the season progresses and the plants get going, there could be snow. In 2020, the first snow fell the Tuesday after Labor Day. The ground held ample heat from weeks of near 100-degree weather, so that sudden dusting hardly fazed the plants. But sometimes everything will die in an intense, early freeze.

I live in USDA Hardiness Zone 5b, sometimes 5a, a designation that threw me for a while. Garden-savvy folk toss around phrases like, "I live in 7a, but these 9b succulents seem to be thriving!" Until I looked for answers, I had only an unsteady grasp on what this meant. I can find it intimidating to hear people apply language in unfamiliar ways, but if I refuse to grow and to learn, I limit myself and the possibilities for the world around me.

Those USDA Hardiness Zones connect to the average lowest winter temperature of a region. They tell a gardener what level of tolerance plants have for cold. The lower the zone number, the lower the winter temperatures. Here in Zone 5a/5b, average lows range from -20 degrees Fahrenheit to -10 degrees. Miami, Florida: Zone 10b. Jackson, Wyoming: Zone 3a. Lead plant, one of the plants recommended for Zone 3, sounds as if it would stand up well to extremely cold winters. To plant citrus trees outside in Colorado would be foolish. They wouldn't make it through a year. But many varieties of apple tree thrive here. Starting in September, I can walk around the neighborhood, seeing tree after tree burdened with more fruit than the yards' owners can handle.

Along with the long-untended clematis bush, some self-seeding coreopsis, and the snapdragons that rose from hardscaped beds, I found a few patches of myrtle spurge when we moved here. With small bright leaflike flowers in spring, myrtle spurge looks a bit like the ice plant that held up Irvine's terraced hills during my childhood. Once a popular low-maintenance ornamental in the Mountain West, the ground cover is now one of Colorado's List A noxious weeds. It is illegal to plant anywhere in the state. Unlike the more self-contained plants I've begun to seek out, myrtle spurge is aggressive in reproductive tendencies and spread.

I learned about the prohibition against myrtle spurge while researching whether to transplant this pretty succulent with the little heart-shaped bracts to other parts of the yard. That's the kind of cautionary tale that might cause distress, for fear of a misstep that could bring on mockery. *You call yourself a gardener and you planted that awful invasive?* Worse, local officials could issue tickets and fines. Residents of Park City, Utah, can be fined up to $100 a day for failing to remove myrtle spurge from their yards. Even the hours of selective weeding needed to remedy such an error can feel like punishment. It seems that any wrong move in the garden could bring on a cascade of calamity. But what is gardening if not a catalog of successes and failures?

I do what I can with the information I know and with the materials I have. And when that information proves incomplete, I seek more. There are plenty of ways to fail in a garden. There are also abundant ways to succeed.

Here's a simple complication: What do I mean when I say the word *nature*?

Even as I build it, my answer shifts. I picture the simultaneously increasing and decreasing heft at the tops of the sand dunes Edward

Abbey describes in *Desert Solitaire*. The instability that is the only stable truth beyond the angle of repose.

When I was a child . . .

Why begin in childhood? As if the chronology of my life matters in some fundamental way. As if linear time is the best path toward constructing a map of life in the world.

Hollyhocks here usually bloom in June, but the hollyhocks that bloomed in September in my dooryard did not bloom out of order. They started when they started. So it is with any story. Even in the Holy Land, where Crusaders first encountered the flowers one thousand years ago and used them to make a salve for their horses' hocks—from which came the name *holy hocks*—the biennial flower bloomed in its own time, if given the space.

I read today—which could be any day, this book spans one year and seven, one lifetime and several—in a book called *Literature for Nonhumans*, that prior to European settlement, wetlands covered three-quarters of the portion of land we now call Illinois.

My father was born in Chicago, Illinois, and raised in Springfield, a town that also claims President Lincoln. Lincoln's grave is in the same cemetery as my grandmother's and uncle's. But my family members are buried near the back of the grounds, in one of the cemetery's two Black sections. Abraham Lincoln's enormous monument has its own pride of place in the thickly treed, meticulously landscaped areas toward the front. Dad says most of Oak Ridge Cemetery resembles a garden, with plenty of trees and flowering shrubs to keep gravesites pleasant and cool. With room for benches where mourners can rest in contemplative quiet. In the cramped Black sections, though, he says, there's no space for benches or trees. "People mourn their dead under the unforgiving midwestern sun."

Long after the Civil War, Reconstruction, and mid-twentieth-century advances in civil rights—well into our ongoing efforts toward a more perfect union—human bodies molder into earth in segregated plots. The men who designed and promoted Oak Ridge Cemetery boasted that they created a final resting place for the people of Springfield "with naught but the pure arch of heaven above us, and Nature in all her silent beauty and loveliness around us." But the construction of this sublime space reinscribed separations informed by prejudice, habitual social practice, and biased allotments of care.

I read in a book called *Literature for Nonhumans* that only 3 percent of Illinois can now be considered wetlands. Though once 75 percent of Illinois (nearly thirty million acres) consisted of land that regularly or periodically flooded. Many plants and animals thrived in the previously well-hydrated terrain. But people redirected rivers and drained land to build towns and cities, roads and farms. Only one million acres remain of the once vast, welcoming wetlands of Illinois. If I traveled there, I imagine finding some of that 3 percent labeled "Nature Reserve" or "Forest Preserve" or "Natural Area." This, then, is one answer to what the word *nature* might mean—someplace separated from the action of everyday lives.

The lexicographers who compiled my *Webster's College Dictionary* included this first definition of nature: "(n.) 1: the natural world as it exists without human beings or civilization." Such a definition allows, even encourages, separation between the interests of people and the interests of the other-than-human beings with whom we share land. European and Euro-American settlers in Illinois prioritized the landscapes of their own childhoods. The drained and cultivated fields and built up towns they understood spelled home. Not the wetlands and prairies they encountered in Illinois or other places as they moved farther and farther west. They altered and degraded the landscape until it conformed to their expectations and demands, ignoring the well-being of any lives but their own.

In the United States, we say *lumber* to mean *timber* because English settlers, who said *lumber* to mean *junk*, cut down vast amounts of timber while clearing land to claim space for the Eastern Seaboard's colonies. They ended up with a surplus of wood they used for construction material: for furniture, for houses, for churches, for courts, for constructing the hulls of sturdy trading/slaving/war ships. Thus evolved the lumber supply store, which sells timber. How efficient. The same ship could drop off a load of West African human cargo in Jamaica, carry slave-harvested sugar to the Carolinas, then pick up Carolina pine planks and pitch as ballast for the journey back to Liverpool, where Englishmen built more ships and more wealth from resources extracted from what seemed, to the Englishmen, *wild* places with endless supplies.

But what's with all this history? This book is supposed to be about my garden.

And it is.

Central Park, one of the masterworks of nineteenth-century landscaping, was designed by Frederick Law Olmsted, who also designed or influenced the designs of the park system in Seattle, the Stanford University campus, where I spent the bulk of my college years, and the Oak Ridge Cemetery in Springfield—the final resting place of Lincoln and of my father's family. All these places impose on nature man's wishes. Olmsted ordered ditches dug, ponds placed, streams routed and rerouted, trees planted for the most ideal views. He even removed people—whole neighborhoods like Seneca Village, a primarily Black and immigrant community of churches and two-story homes and a school—from what became Central Park. Olmsted fabricated nature to make it look particularly "natural."

Although invited, Olmsted refused to take on the Golden Gate Park project in San Francisco. The giant sand dunes, still called "The Outside Lands," on the section of the city that became the park confounded him. Olmsted tamed natural landscapes, brought them within his control, and that particular western landscape seemed, to him, intractable.

Thinking of the wild, whipping hollyhock in my dooryard, I smile.

I love the intractable bits.

"Don't you want some gloves?" Mom asked when she came by the house and found me digging in the central flower bed.

Unless I'm handling plants like the myrtle spurge, whose glue-white sap raises rashes on uncovered skin, I prefer to work with bare hands. Even writhing worms no longer repulse me.

Science provides reasons for the pleasure I feel when I dig with my hands. Microbes in the soil lower stress responses, raise serotonin levels, and may reduce inflammation in the body and brain. But when I dig, the science is not what I think about. I think about how good it feels to welcome growth.

Late-September days are perfect for putting in arrays of tulip and crocus bulbs here, tiny round black columbine seeds there. I spend warm afternoon hours on my hands and knees, digging in the soil. Some seeds I bury half an inch deep, protecting them from sharp-eyed birds. I scatter others over the surface the way wind gusts scatter dry pods' offerings. Bulbs prefer various depths as well. Crocuses, tulips, alliums, and grape hyacinth: four inches. Midspring daffodils: five. Summer-blooming bearded iris: two. Savvy gardeners layer varieties of bulbs and seeds to enjoy continuous blooms.

In late summer and through the fall, it's easy for me to duck into a nursery for a plant stake or trowel, walk past racks of seeds and bins of bulbs with their bright illustrations of future flowerheads, and find fifteen dollars' worth of seed packets and bulbs in my handbasket by the time I reach the checkout. And it's free to collect seeds from my yard and from friends. Cosmo seeds, like tiny pencil leads. Sunflower seeds in all sizes, looking just like the ones people eat.

The big stand of sunflowers in our front yard sometimes spills over the sidewalk. I should prune the seedy three- and four-inch-diameter deadheads so they don't block the walkway as they pull the plants down. After helping Callie with a math assignment and putting dinner in the crockpot one September afternoon, I approached the sunflowers, shears in hand. A flock of pine siskins rose from the stalks. To be honest, I didn't learn they were pine siskins until the encounter drove me to our bird books. I had noticed only goldfinches around the sunflowers before. Less showy than their goldfinch cousins, and no bigger than goose eggs, these little pine siskins have streaked-brown plumage that blends into the browning stand of autumn sunflower stalks.

One stayed to assess how close I might come. Saving the energy others spent on fleeing.

"It's okay, little bird," I whispered. "Keep eating."

As soon as I stepped away, the flock returned to the seed-rich sunflower bed.

I didn't know, when we chose not to use chemicals in our yard, that glyphosate, one of the main components in the weed killers the former residents preferred, disrupts serotonin uptake in animals, thus suppressing the beneficial powers of getting my hands dirty. I knew only that much of what I want to grow would be considered, by those herbicides, to be some kind of weed. Purple prairie clover, whose pen-cap-size tubular heads are ringed in seeds small as the dots on strawberries. Prairie dropseed, a mounding native grass with bursts of panicles—branched clusters of flowers from which drop the plant's popcorn-scented seed. Monarda, sometimes called bee balm, draws many flying creatures in addition to bees. It makes me happy to see these plants coming up, sometimes willy-nilly, throughout the yard. I am happier still encountering the community of life that visits. The long-horned sunflower bee (*Svastra obliqua*) that looks like a stretched-out European honeybee with extra-long antennae. The white-shouldered bumblebee. The plainer *Bombus nevadensis* and *Bombus occidentalis*. Also *Bombus centralis* and *Bombus huntii*—bombastic black-, yellow-, and orange-striped native bumblebees. Goldfinches, dragonflies, rabbits, pine siskins. Learning all these names took me years. Learning a name for the joy of this grounding may take a lifetime.

Many of the plants that grow in what might have once been called the wilderness of my garden are scraggily. Rangy and wild. Some have furlike thorns that make them more difficult to manage than the most touch-resistant rose. If they carry a scent, it might be the scent of skunk or a cud of brittle grass. Their little flowers die the moment I cut them from the stem, rendering them useless for the vase. *Weed*, from the Old Saxon *weód*: "a useless or injurious plant."

One fall, trying to gather up all the beauty I could before an early freeze, I cut a bouquet of snapdragons only to find the small, dragonlike mouths of every flower filled with tiny mites. Until I cut the flowers, I'd never noticed these mites. Now, before I share the tiny black snapdragon seeds with friends or scatter them in my parents' garden or my own, I empty the dried pods over a mesh screen to sift out the minuscule bugs.

A delicate larkspur, called Nuttall's larkspur, grows in clusters near the snapdragon. I love that little flower. Several bluish, purplish starfish, each with a dark tail, like a spur, bob off the plants' light green branches. All winter, I look forward to clusters of these blue-purple flowers returning to patches around the yard in the spring.

I don't keep cattle, to whom Nuttall's larkspur can be toxic. This helps explain why I am not bothered to see it on my land. I would rather foster the larkspur's growth than treat it as an injurious plant. Cattle ranchers also list as undesirable the mint-green shrubby stalks of rabbitbrush I cultivate in our yard. And the low-growing white flower some call locoweed, which is said to drive cattle crazy. Also swamp and showy and western whorled milkweeds, with pods that split and scatter cotton-tailed seeds far and wide. Though they are native to this landscape, these plants interfere with commerce and often show up on lists of undesirable weeds. Walking her daughter to my door, the mother of one of Callie's friends stopped near our milkweed's rowdy pinkish flowerheads. "I got rid of so much of that when I was a kid on the farm," the woman said.

I showed my father our brilliant yellow-flowered rabbitbrush once, pouting about seeing the plant listed in a book called *Weeds of the West*—a catalog that also sweeps into its pages Nuttall's larkspur, locoweed, and all those varieties of milkweed.

"A weed," Dad said, "is a plant that is growing in a place or a way you don't want it to grow. That's all that word means."

Along the rocky margins near our house's south wall grow plants that stick to my clothes like the burrs in the hills of my childhood. I find them troublesome and unattractive. One grass, hare barley, is nearly impossible to extract once it catches a hem. The needle-sharp sections of a mature plant's two-inch bristly spikes break off easily and stick to anything they touch. According to *Weeds of the West*, this includes the soft mouths and nostrils of grazing livestock. Year after year, I pull hare barley from the rocky section where it grows. But, as I prefer to concentrate on what I want to see growing than what I want to see gone, my efforts are halfhearted, and the hare barely never entirely disappears.

The page in the weed book that listed Rocky Mountain iris reinforced my hesitancy to call a weed a weed. Rocky Mountain iris is indigenous to this region. It comes up in gardens as well as along trails. Rising from weathered soil like a purple flag—some call it the blue flag iris—its white- and yellow-tongued mouth opens along with my rising excitement for spring and summer's blooms.

Synonyms for *wild* include: *natural, undomesticated, savage, desolate, uncultivated, unbroken, uncontrolled, impractical, disorderly, rowdy, ill advised, waste.* Blue flag iris doesn't work well in bouquets. Cattle shouldn't eat them. Their blooms don't last long in the garden. Still, I welcome the blue flag irises' ephemeral upheaval each June. I cheer when their blades push aside soil.

Once, when Ray and I still lived in Northern California, we visited John Muir's house specifically to take a tour with a particular ranger. I had a bit of a crush on this ranger, who was smart and dark and kind and tall and funny and shared a name with my own smart, dark, kind, tall, funny husband. Imagine encountering a beautiful flowering plant and then visiting the places it grows again and again, never with the desire to cut its bloom to keep for yourself. I simply enjoyed hearing this ranger talk. Mind crushes I call these, when I find myself loving the shapes and movements of another thinker's brain. One way I experience love is as delight in the flowering of another's ideas.

I want this book to offer such a flowering.

The John Muir house is lovely. A National Park Service site in Martinez, the ten-thousand-square-foot Italianate mansion is large even by today's standards. Airy, with seventeen spacious rooms, it boasts an impressive office where Muir wrote after his tramps through the Sierra, to Alaska, across the country, and into the boardrooms of powerful men. Muir's father-in-law was a doctor and kept his clinic off the side of the house. Medicine bottles spun of blue and green glass sparkled in the light that streamed through the windows.

Around Muir's house—truly, this was Muir's father-in-law's house, but with fame frequently comes the power of possession—hills rustled in golden glory. Dried out in the long months since the rains, grasses bowed in gentle wind. I found myself looking out the house's windows. What a beautiful situation! I wouldn't mind living there. But, when our tour ended, rangers escorted us out of the building to make room for the next scheduled group.

Before driving home, we stopped for a picnic in Muir's expansive yard.

On one of my bookshelves here in Colorado, I keep a brick from a rubble pile in an area of Richmond called Shockoe Bottom. I gathered it back when I lived and taught in Virginia. My friend Vanessa, who is now an American history professor, says that the warehouse district where I picked up the brick was once known as the Devil's Half Acre. For more than three decades, enslavers penned as many as 350,000 Black men, women, and children in Shockoe Bottom. From 1830 to 1865, Richmond made itself the United States' second-largest site for the sale of enslaved people—people off of whose bodies and lives enslavers grew rich. In the years after the abolition of the legally sanctioned international slave trade and the invention of the cotton gin, those Richmond warehouses trafficked hundreds of thousands of human beings. America's economy grew more and more ravenous for the labor and lives of Black people born on (some would say *from*, and some would say *for*) this soil.

Sometime after I picked up the brick, investors redeveloped that part of Richmond into condos and shops for wealthy Virginians. Even when I walked there, Shockoe Bottom's past was rewritten to honor the warehousing of tobacco, not the enslaved. I keep that brick on my bookshelf to remind me of history's baked-in, often horrific, and frequently overlooked layers. Beside the brick—perhaps by happenstance, perhaps by design—I placed the pecan from John Muir's orchards.

The afternoon we visited Muir's property, a friend—one of the readers from our wedding—walked with us through the tree rows. He ate several pecans. Pocketed a few more to take home. I remember thinking that my friend displayed a particularly white male sense of entitlement. I marveled at his unfettered gumption, his wholly unselfconscious assurance of his right to partake of what the orchard produced. I, surreptitiously, picked up only the one, loving its pointed and shiny brown shell. When Ray and Callie and I moved from California to Colorado, I carefully wrapped the pecan and brought it along.

The national park set aside a large room dedicated to displaying Muir's work: the books and articles and letters he wrote, the organizations he founded, the land he helped preserve. Muir was a great man, these displays maintained. He successfully fought to establish and protect Yosemite, Sequoia, and Grand Canyon, some of the first US national parks. He helped found the Sierra Club in 1892, serving as its president until his death twenty-two years later. He devised a theory of glaciation that guided scientific and geological research for more than a century. His writing helped map this nation's imagination of its wilderness and how that wilderness should be managed. The displays in the museum suggested that I should be thankful for what John Muir left behind. But I wish I could have heard his wife, Louie, play her piano. I wish I could have tasted the cook's food. What of Muir's two daughters? Did they sing with their mother's music? I want the inside of Muir's house—the people there, *that* environment—also represented in his stories about the world he worked so hard to preserve. But that's not what Muir thought people wanted. Or, perhaps, he didn't consider those subjects worthy of his time and quill.

I am angry at Muir for what he left off his pages. Angry about the dismissive and degrading ideas about women, and Black people, and Latinx and Native folks he included.

I don't believe John Muir or the foundational environmentalists of the nineteenth and twentieth centuries had much interest in me. I mean that in terms of so many of the people who live inside me. Muir took Louie, the mother of his two children, to Yosemite once, but because her pace differed from his, and her fears and fascinations remained independent of his own, he grew annoyed. He never brought her along on his expeditions again. I don't have the kind of body, or the kind of money, to give me access to the sort of power John Muir preferred. If I have skin in the game men like Muir constructed, it's the wrong kind of skin.

The organizations Muir helped create got the ear of the most powerful men in the country because they spoke to what these men wanted to see in themselves. They spoke to their ideas about suitable forms of recreation—recreation that accented their power or gave them safe space to relax. They excluded those who weren't wealthy white men—extending this exclusion to Indigenous communities who had long lived in and thoughtfully tended what the 1964 Wilderness Act refers to as land "untrammeled by man." It's a fantasy, this vision of wilderness. A fantasy designed to serve only a limited few.

Maintaining the fantasy of the American Wilderness requires a great deal of work. It requires the enforced silence of women, of Black people, Chinese people, Japanese people, other East and South Asian communities, poorer white people, Indigenous people, Latinx people, human children, wolf cubs, other small and large mammals, lives that thrived in wetlands, lives that thrived in grassland prairies, lives that thrived in the desert, flower people, fish people, bird people—the list goes on and on.

One of the reasons so many Black gardeners expressed such fear on those online conversation boards is that this fantasy pervades so much American thinking it is sometimes hard to recognize the exclusionary vision for what it is. People who have been erased from the vision feel like they are trespassing in territory where they're not allowed. Or like their presence will be condoned in only a qualified manner. A conscious or unconscious awareness that one's claim on the world could be revoked at any moment leaves no room for accidents or explorations. No room for planting a plant that won't grow "properly." Or for turning down a dead-end path. No room to feel comfortable gleaning a fallen pecan.

I often hear the catchphrase "representation matters" connected to books and movies and television shows that actively portray the lives of systemically marginalized people. If self-actualization means being able to look in the mirror and (positively) recognize myself, it also demands that I be fully myself, regardless of systems designed to erase me. I wrote about my crush on the ranger because I want to be clear that I went to Martinez to hear a Black man I admire. I didn't go there on a pilgrimage to honor John Muir. I wrote about that ranger because, even in Muir's house, in that space of calculated exclusion, that ranger welcomed me.

My daughter's days are divided. She moves through subjects—reading, math, social studies, and the next day she does some version of the same. Fall, winter, spring, and again fall: the same story with slight variations. It's a kind of indoctrination, this brand of education. Drilling into children that some voices are for outside and others for inside the house. But, when most effective, learning leaves room for reframing. Overhearing and overseeing her lessons as Callie, in the house alongside me, learned fundamental building blocks of the knowledge sets that construct our world, challenged me to think differently about what I've learned and what I've had to discover. And so, 2020 turned into a year for relearning—or, more accurately, a year to discover how to better see truths that have been present all along.

"We learned about Manifest Destiny today," Callie told us at the dinner table.

She said the words *Manifest Destiny* with such disdain that I doubted her lessons about westward expansion were as romanticized as mine had been when *Little House on the Prairie* was one of the country's most popular TV shows.

"What did you learn?" I asked.

"People needed to conquer all the land, even though the land was already owned. Well, not really owned, but lived on. White people thought they wanted all the land. The Native Americans said, 'No, thank you,' but the white people wanted it anyway, so they bombed their homes to get the land they wanted."

Ray and I put down our forks. "They bombed them?"

"Well, they didn't really use bombs. But they took the land."

"You learned this in school?"

"We learned about Manifest Destiny. We learned about them shooting Native Americans."

"Where'd you learn the rest?" Ray asked.

Callie looked pointedly at Ray and at me.

The 1804–6 Corps of Discovery Expedition opened the West to a flow of settlers driven by ideas of Manifest Destiny. The white men who trekked across the continent with Meriwether Lewis and William Clark received enormous land grants. With these grants came the perceived rights to do with that land what they wished. I could write a book on what this means to us in the first decades of the twenty-first century. In a way, this is that book. Though native to the place the white men on the expedition claimed to discover, and though without her knowledge and experience the white men would have been lost, Sacagawea received no land grant. Upon the successful completion of their journey, York, the Black enslaved man who, by many reports, did far more than one man's share of work on the journey, didn't even receive the manumission Clark promised. The vast wilderness Thomas Jefferson's commission claimed was intended only for white men. As Callie put it, these men "wanted all the land."

So much of this nation's environmental vision descends from a commission intended to benefit only a select few. But by naming the violence that ensured "westward expansion"—violence that encouraged a fantasy that much of the West was "untrammeled by man"—Callie joined the work of claiming space for herself, and for others whom such violence works to erase.

Like the hollyhocks that grow more audacious when they detect a chill in the air, like the native wildflowers I admire, survivors made heartier by drought years—blooming after, often because of, hot fires, hard frosts—my family and I have set down roots in Colorado.

Someone asked me yesterday what hope looks like. Yesterday could be any day. The hollyhocks had not bloomed, were blooming, were already spent. Tomorrow began years ago. When, just yesterday, someone asked me what hope looks like, they expected an answer that had something to do with protests, elections, and classroom pedagogy. But I'd just been outside. Even inside, the air from outside flowed through my fabric, close to my skin.

"My garden," I answered, recalling the pine siskins rustling in the sunflowers. The bulbs I plant four to six inches deep every fall, whose blooms I believe in, though they won't manifest for months to come.

The wildly whipping hollyhocks are welcome here, and so are the little purple larkspurs, the snapdragons, the birds, and the bugs, and the rabbits. I don't know who everyone is yet or where they will stand with my neighbors and me. But I am learning how to see, and how to count them, hopefully, more abundantly.

On a break from her schoolwork when summer chilled into autumn, Callie and I placed a new birdbath in the side yard where the prairie project rooted. Callie arranged two little piles of river rocks that jutted out of the water so bees could land and drink. A few days later, from the window near our piano, I watched a couple buzz a circle around the new birdbath and settle onto the rocky rest Callie so carefully built. The bees stayed there, safely—long enough to take a satisfying drink.

Hollyhock with pollen

Clearing

all night the wind blows   & my mind
    my mind is like the hawthorn that loses
limbs   they litter the ground   crush
    the black-eyed susan   scatter buds
over rows of lettuce   bean sprouts
    whose greens are clusters of worry
in raised beds   blown leaves & cracked limbs
    threaten our foundation   water backs up
in gutters   seeps into the house's walls

    but my mind   my mind is not in the house

in the yard's far corner the eye of my mind rests
    on a hawthorn branch   shaken   snapping
hectic   then still   the day dawns
    without anger   the blue jay I've looked for
pushes sky off his crest   how splendid
    his wings & tail   it's not so much
that before this he'd hidden himself
    it's only he favored a roost
I could not see until the storm thinned the tree

Callie waters a potted Zinnia haageana *plant*

When Callie was six, we visited neighbors in their back-yard. Callie beelined for their trampoline to start jumping. The neighbors' boys were grown and gone. In their absence, a honey locust drooped low over the bounce pad. Callie grabbed its limbs and tried to keep hold as she dropped close to the ground. Two branches broke with a sound like the snap of a carrot.

"Don't hurt that tree," I said.

Without thinking, I added, "Trees are people too."

The conversation stopped.

"Did you just say that trees are people?" asked Ray.

I sometimes teach environmental literature classes at the university. When I do, I show students a sixteenth-century copperplate engraving known as *The Great Chain of Being*. It's an excerpt from the series *Rhetorica Christiana*, created by Diego de Valadés in 1579. God reigns from a cloud-borne throne around which shines an aura brighter than the sun's. He's bearded, with rich robes and a head-piece as tall as a three-year-old child. Jesus slumps on God's lap, arms spread as if he's only recently come off the cross, his upper body exposed, and his lower body modestly shrouded. He's a rather European-looking son of God. This artist, even this artist, born in what's now northern Mexico to a Spanish conquistador father and a Tlaxcaltec mother, apparently found it easier to imagine winged angels flying near the Lord than to imagine a Brown or Semitic person leaning into such power, touching God.

A heavenly scene—God, His son, the angels, the great clouds of glory—dominates the upper section of the image. Given the scale of the rest of the beings depicted, Jesus and God are, as the saying goes, larger than life. Below God and His son, a row of worshipful angels stands on a layer of cloud. Below that, and smaller, men and women walk on the firmament, heads uplifted to God. Below that again, birds get their own row. In class, we identify a peacock, a pheasant, a goose, a duck, a hawk, a quail—on branches and in the air. Carry on farther down the engraving to a row with water and marine lives of all kinds: a grouper, a salmon, a squid, a clam, a lobster, a crab, a snail. Valadés engraved this copperplate two centuries before Carl Linnaeus created the binomial nomenclature we use to identify animals and plants. But in this depiction of *The Great Chain of Being*, a clear taxonomical system is already in place. Living beings are ordered, classified, and divided.

On a less sturdy strip of land, three rows below the one reserved for humans, Valadés depicted real and fantastic animals: a camel, a rhinoceros, a unicorn, a horse, an elephant, a dog, a goat, a boar, a dragon, a hyena, a monkey, a mouse, a stag. Below the beasts, nearly at the base of it all, planted into rolling hills: pine trees and palm trees and willows and firs and ash trees and oaks and elms. Twenty specimens to represent all the flora of the world. According to the taxonomy of value inscribed in *The Great Chain of Being*, trees are very far from being people.

The Great Chain of Being is one of twenty-seven plates intended for Franciscan missionaries as they converted Indigenous people in the so-called New World. If their dark converts could be taught the ways and the words and the rites of the faith, then, unlike beasts who could

An excerpt from Rhetorica Christiana,
created by Diego de Valadés in 1579

not be converted, they must be capable of salvation and the grace God extends to humanity. In this way, Diego de Valadés worked to prove that his mother's people were no less worthy of God's love than his father's. *My God*, I thought, when I read the why and wherefore of the engraving of *The Great Chain of Being*. This image sums up much of the thinking that still drives so many interactions in the world.

Back at the top of the engraving, surrounding God and dividing Him from all that's below, fly a well-spaced ring of seven exultant archangels, enshrouded in ethereal clouds. Mary, or someone who looks like an icon of the Virgin Mary, floats inside the ring, set off to the side by herself. Radiant, fully clothed, already finished with the immaculate conception and the most holy birth, Mary kneels on a cloud, deep in prayer. Her isolation seems part of her holiness.

At the center of the ring of angels, God holds a chain in one hand. We can follow its links down all the levels, to the bottom of the engraving—in this worldview, there is clearly a top and a bottom—where the chain's final link is affixed to the devil's head. The naked devil has breasts like a mature woman's, like my own, sagging as if they had once been heavy with milk. The devil's legs are bent and spread open, as if she's seated on a birthing chair. Near her stomach and groin may be rolls of fat in the shape of a mutated smile, but I think I see a newly crowned baby. As if this devil is constantly giving birth to new demons. As if a woman visibly bearing a child must be suspect. Arrayed around the devil, at the bottom of the chain, a tumultuous gang of lesser demons causes a ruckus, flogging and flailing, stirring up other demons in pits of despair. It looks like hell down there.

When I teach this image, someone in my classes will eventually point to a winged angel on the right column of the copperplate. A fallen angel captured in the process of descending into hell. That's what you get, the engraving seems to want to remind us, when you mess with prescribed hierarchies of power.

To believe that the way of our world is the natural order of a prescribed chain of being is to believe one way of living in the world is worthier than other ways. This makes it easier to justify brutal demands on and dismissals of lives that are different.

There *are* other ways to see the world than through the rigidly segregated framework conveyed by *The Great Chain of Being*. Even Saint Francis—after whom Diego de Valadés's religious order, the Franciscans, took their name—believed God was present throughout all creation. I want to believe it is possible, even necessary, to think of the world not in a top-down, divided system, but in an interconnected way. But life grows more complex when we muddle hierarchies of power.

Sometime in August of our homebound year, before the hollyhocks in our dooryard bloomed, but after the sunflowers grew past Ray's head, I reread Bill McKibben's *The End of Nature*. McKibben first published his treatise on the end of the natural world in 1989, when I was sixteen and didn't want to think that the world I'd hardly started living in was already destroyed. More than three decades later, I found McKibben's book no less terrifying. Among other things, McKibben warned of the irreversible devastation awaiting the planet if the carbon dioxide in Earth's atmosphere exceeded 350 parts per million (ppm). In December 2020, we reached 412.5 ppm.

When I say, "We reached 412.5 ppm," I mean all who share this planet. Every living being experiences the repercussions, though some feel the effects sooner and more violently, while a small group of humans—I am included—dole out much more of the damage.

Some causes of our high carbon dioxide (and methane) output: the burning and removal of trees and native vegetation; dependence on corn and soy, which, like hunger for bacon and beef, bears part of the blame for the burning and removal of native vegetation; the melting of ice caps; the impoverishment of soil; the consumption of coal; the unending extraction and expenditure of oil and natural gas (humans had not yet accessed the tar sands when McKibben first published *The End of Nature*; humans had not yet fracked shale in Texas's Permian Basin and in Pennsylvania and Oklahoma and Colorado). "Global warming is not a problem for the future," wrote McKibben, who was not the first, nor the last, to sound this alarm. "It is here now." One thing that makes *The End of Nature* so terrifying is that McKibben is correct about what was and is and will be happening because of human actions and their repercussions. But he underestimated our intransigence, and things are already much worse than his book imagined.

McKibben knew Earth's atmospheric carbon dioxide saturation had risen at the unprecedented rate of about 2 ppm per year since 1970, when, according to the Global Monitoring Laboratory (GML) on the north side of the Hawaiian mountain Mauna Loa, levels registered at 325.68 ppm. Since the initial publication of McKibben's book, the growth rate of atmospheric carbon dioxide saturation is ever increasing. The GML on Mauna Loa registered 411.29 ppm in September 2020, up from 408.54 in September 2019 but likely well *below* the level of carbon dioxide saturation in the atmosphere when you read this.

A citizen-led initiative called CO_2 Earth compiles and publishes global emissions and temperature data, including the GML measurements. On its website, CO_2 Earth writes, "The world cannot stabilize what it does not watch." It seems important to see these numbers and begin to know what they mean. For the same reason, I want my students to acknowledge belief systems that may consciously and unconsciously drive actions. Belief systems like those inscribed in the Valadés copperplate. Without interrogating our values, our actions will be impossible to change.

In *The End of Nature*, McKibben says Saint Francis believed "just as God had sent Jesus to manifest him in human form, so too he represented himself in birds and flowers, streams and boulders, sun and moon, the sweetness of the air." McKibben suggests that, according to this saint of Christian tradition, nature can be "a way to recognize God and to talk about who he is."

If I understand God as separate, as *above* all creation, then what happens elsewhere, to others, may not matter much to me. But let me believe God is *in* all creation, that birds and beasts and boulders and streams are all part of God's body. How much better might I treat the lives around me?

"Don't hurt that tree. Trees are people too."

I was born into a season of water restriction, a child of 1970s Colorado and Southern California. Fuel-efficient cars sported bumper stickers that normalized water conservation with sayings like SAVE WATER, SHOWER WITH A FRIEND. I've heard people from all over the United States say, "God's not making more land." As a child in the West, I also learned God's not making more water.

The Colorado River, whose headwaters are less than fifty miles from our house, has made it all the way to the Gulf of California only once in this new millennium. It rarely even reaches its estuary in northwest Mexico. Before all the dams and diversions along the river's course, that estuary, the Colorado River delta, was among the world's lushest and most biodiverse regions. Halfway through her 2012 essay, "Wager for Rain," Megan Kimble writes, "In 1922, when the Colorado River Compact was signed to divvy up the water of its namesake, hydrologists measured the Colorado River at an annual flow of 16.4 million acre-feet." The problem was, they measured during one of the wettest years on record. The typical flow in the early to mid-twentieth century would have measured 3.5 million acre-feet less. Even that reduced volume seems optimistic in the twenty-first century. Between 2005 and 2020, the flow rate along the 1,450-mile river—a river that supplies water to more than forty million humans in Colorado, Wyoming, Utah, New Mexico, Arizona, Nevada, California, and sometimes Mexico—shrank to closer to 12.5 million acre-feet.

We connected an eighty-five-gallon rain barrel to a downspout at the rear of our house. Six feet tall and shaped like a terra-cotta urn, the barrel captures snowmelt that spills off our roof and the rain that runs off during the two-month monsoon season that historically quenched this dry state. According to Colorado House Bill 16-1005, we can collect a maximum of 110 gallons of water in such containers. What our barrel harvests must be used in landscaping beds or on the lawn. We can't keep the water for household purposes like laundry or showers. Even after filtration, using captured precipitation for human consumption is illegal. According to a Colorado State University Extension fact sheet: "In our arid environment, every drop counts, and water rights holders depend upon the runoff from snowmelt and rainfall to supply the beneficial uses to which they apply their water rights." This includes the agricultural needs of all states that lay claim to the river; the people (and livestock) fed by the fruits of those agricultural basins; the towns and cities in which millions have settled; and, though they tend to be the lowest-priority claimants, the plants and animals that have relied on the river's hydration for eons. "Captured precipitation that is consumed 'out of priority,'" the fact sheet continues, "may deprive downstream and/or senior water rights holders of their right to use water from the natural stream." As my family uses water here at the upper portion of the Colorado River's usage corridor, we alter the condition of millions—billions—of lives.

When Callie was four, I walked into the bathroom to find her running water from the faucet. Just standing there watching good water disappear down the drain.

"You'll be sorry you wasted all that water when the water wars come," I said.

Living through perpetual drought has made me a bummer of a mother.

She turned off the tap.

"What's that face you're making?" I asked. Callie looked in the mirror with a fierce expression I'd never seen before.

"This is my water wars face," my girl scowled.

Desert as a synonym for *wilderness* suggests a godforsaken place with no hope of help or water. In stories, most people will do most anything not to wander the desert unaided. The West's water scarcity, and arguments over who has a right to water and who doesn't, have been among the fundamental constructs of my life.

When I was still a baby living not far from where my daughter practiced her water wars face, Eudora Welty published a review of *Pilgrim at Tinker Creek* in the *New York Times*. Welty began the review with a quote from Annie Dillard: "I am no scientist, but a poet and a walker with a background in theology and a penchant for quirky facts." Forgiving of but not wholly convinced by Dillard's meandering mind and methods, Welty's assessment of the book is rather icy. Welty seemed to see Dillard's writing as ambitious, but somewhat naive. For her part, Dillard says her book wrestles with God's possible manifestations. An ambitious pursuit, certainly, and one as unlikely to yield permanent clarity as the Colorado River is liable to slake all of our thirsts.

Dillard prefers not to think of *Pilgrim at Tinker Creek* as a meditation on nature. Rather, she considers it a theological treatise. And so perhaps we have in her book an example of the call to see God in all creation. I appreciate that. I have looked for such connection all my life. Part of my drive to think so deeply about the greater-than-human world in direct relationship with my personal and cultural history comes from a desire to construct meaning from and connection with what is beyond me *and also* what binds me to the rest of the world. It's a spiritual question—and a practical one.

Harriet Tubman relied on signs and signals communicated by owls and snakes and moss and river currents. Such connections helped her safely conduct herself and others through a dangerous landscape. One made more dangerous by humans—made *far* more dangerous by humans—than by nature's wild predations. Dr. George Washington Carver put words to the kind of radical connection on which Tubman relied: "I love to think of nature as an unlimited broadcasting station, through which God speaks to us every hour, if we will only tune in." From 1896 until his death in 1943, Carver taught botany and agriculture at Tuskegee Normal and Industrial Institute, a teaching college and trade school founded to provide Black students with ample practical opportunities to succeed in the consistently hostile environment into which they would graduate. Carver spent time every day in his garden or walking the woods near his house and the campus. My granddad Dungy studied at Tuskegee in the 1920s. It's a small world, and Tuskegee is a small campus. I hope my grandfather and Carver walked together sometimes.

"Acts of creation are ordinarily reserved for gods and poets, but humbler folk may circumvent this restriction if they know how," wrote Aldo Leopold, a man considered by many to be the father of American wildlife management. "To plant a pine, for example, one need be neither god nor poet; one need only own a shovel."

I've never restored a white pine forest as Leopold did, but, like him, I've spent a large portion of my life trying to think like the landscape. I want to access Earth's splendor while benefiting the flora and fauna around (not below) me.

In the environmental canon, there are plenty of other examples of such reaching toward God, or whatever name can be given to the vast and powerful energy force binding creation—books in which the faces of God are the faces of bears and muskrats and white pine and fire and wind and water. "God's love is manifest in the landscape as in a face," wrote John Muir.

Seeking the many manifestations of God, I plant restorative love in my garden—and in this book.

The Indigenous Americans whom Franciscans campaigned to convert likely believed their gods were in everything. But the missionaries accepted only one God—a God separate and above—and linked their faith to this hierarchy of power. Still, like Diego Valadés, I am a child of both Europe's Judeo-Christian traditions and "the New World." I look for the face of God in animals, vegetables, and minerals too. When I lived in Lynchburg, surrounded by a religious intensity that demanded I explain why I bucked the faith traditions of my grandfather by not regularly spending Sunday mornings in the old brick church, I answered: *I prefer to worship at the Church of God in the Great Outdoors.*

I didn't read the 1989 version of *The End of Nature*, but a revised edition published in 2006. In some ways of thinking, that's still a long time ago. What is time but a record of cycles of violence? Flood year, drought year, wildfire, health crisis, war. There have been many of all of those since 2006. When she practiced her water wars face, Callie understood she needed to prepare for more. "We are no longer able to think of ourselves as a species tossed about by larger forces," McKibben wrote in the revised book. "Now we *are* those larger forces. That was what I meant by the 'end of nature.' " In a world where humans alter weather and fire and genomes and the atmosphere's chemicals and the output of rivers and what grows and what does not grow and when, McKibben wondered, do we still need God?

I am no angel. I struggle daily with what I eat and what I drink and how much water I use and how the products of my life hinder our ability to build a more equitable and sustainable world. If I reduce my dependence on factory-farmed dairy, the production of my almond milk contributes to the disruption of traditional salmon runs by requiring the diversion of 1.3 million gallons of water per acre. If I substitute soy for beef and pork, how many tons of pesticides and herbicides are sprayed on soybean fields for me? If I eat hunted elk and deer, perhaps I prevent dangerous population irruptions once kept in check by the wolves and big cats that humans removed from this region. Or perhaps I do nothing but enjoy a delicious meal.

I am still embarrassed by a conversation I had in college with a lifelong vegetarian. For years, I didn't eat land animals. I could understand not eating mammals, not eating fowl. I could even understand not eating octopus. "Octopus are smart," I told my friend. But I couldn't understand why she wouldn't eat fish. "Fish," I said, "are dumb."

By what standard had I measured intelligence?

Describing fish intelligence for *Australian Geographic*, Peter Meredith, a science writer, says we "tend to equate ancient with primitive, clinging to an obsolete view of evolution as a steady progression from inferior to advanced, with highly intelligent humans at the pinnacle. But that's turning out to be a highly simplistic reading of facts." When I called fish dumb, I enacted the kind of hierarchical thinking displayed in *The Great Chain of Being*. My assessment of intelligence dictated what lives I did and did not value.

In the Bible I was raised reading, God created the earth and the heavens and the oceans and man and woman and all the plants and beasts of the land and the sky in six days. Fast, by my reckoning, even if, as Muir wrote, "with God, a thousand years is as a day." Perhaps God's need for speed in the creation is the root of the desire for quick results that entices me to use heavy machinery to clear the rock from my yard rather than my slow, manual method. As if it might be a moral imperative to finish a job as quickly as I can.

First, God divided dark from light and day from night. Then, He set apart the sky from the clouds from the watery planet. Always a He in the Bible I read, God was a man of unquestionable power. On the third day, He separated the water from the land. On the land, He made manifest plants that bear fruit. Grains and trees and vines and bushes. It takes longer than one day to grow plants now, but, my God, He kept busy. On the fourth day, God created the way to keep track of the timing for festivals and holy days during which we could praise Him. Then He created the sun, moon, and stars.

On the fifth day, God looked to the skies and to the waters. The men who compiled the Bible report that He created fish to fill the waters and monsters for the seas as well. Maybe monsters means right whales and sperm whales and blue whales and narwhals. Maybe the walrus, mammal though the walrus also is. Maybe the colossal squid. God populated the waters, then He turned to the sky. He created eagle, albatross, swift, and wren, and lark, and bunting. I suppose this means He created blue jay and cormorant and gull and cuckoo and magpie and harlequin duck and nuthatch and owl. In all likelihood, He created the dodo and the ostrich and the emu and the penguin on the fifth day, though their placement with the birds of the sky is complicated by their inability to fly.

Then came the sixth day, when God made the beasts of the land. Assuming He already made manifest the elephant seal and the green turtle and the painted turtle and the puffin and the western chorus frog and all the other beasts who blend between land and water, and assuming He already made manifest the flightless birds, now God made the lion and the sheep and the boar and the bear and the rhinoceros and the goat and the pronghorn and the oryx and the bison and the buffalo and the deer and the moose and the tortoise and the prairie dog and the house cat and the bobcat and the porcupine and the pangolin and the lizard. The Bible I grew up reading never mentioned butterflies or wasps or mosquitoes in accounts of the Creation, so I don't know if they came out of water, air, or land. But on this sixth day, the Bible says, God created all the animals, domestic and wild, and so at some point He would have had to make the digger bee and the hoverfly along with the black-footed ferret, and the shrew, and both the long-tailed and the short-tailed weasel. As I said, He was busy, my God.

But it wasn't separating the swirling mass of darkness from

the firmament, or dividing land from sky from water. It wasn't the manifestation of plant life or of fish or of fowl, or bringing into being the beasts of the land. It's what happened next that made the difference. What happened next set into motion the beginning and the end of the world. God made human beings in His image. To rule over all the beasts of the land and the sky and the water. To have dominion over everything. God made human beings, as the Good News Translation puts it, to "live over all the Earth and bring it under their control." That was the sixth day. After which, the Bible claims, God rested.

In more than one way, relinquishing control of my garden is antithetical to the constructs in which I was raised. The Bible I was raised reading—now I'm quoting the New Revised Standard Version—says God created man to "fill the earth and subdue it, and have dominion over the fish of the sea and over the birds of the heavens and over every living thing that moves on the earth." That is written in the very first book, the twenty-eighth verse of the very first chapter, that grant of inordinate power.

The sweeping lawns and pristine gardens ushered into popularity by seventeenth-century noblemen and landed gentry are offshoots of *The Great Chain of Being* and the twenty-eighth verse of the Bible. God over all—then the angels, then man, and, nearly at the bottom, hardly differentiable from the inert land itself, the bushes and trees, which are man's to command. Europe's wealthiest men and women designed vistas to view through their many windows. Such views implied far-reaching domain. The wealthy strolled at leisure over manicured pathways and lawns. They didn't get their own hands dirty. Teams of more lowly men and women did the planting and trimming and pruning and mowing and raking needed to keep everything bountiful, lovely, and pristine.

The culture I grew up inside insists I should control what's seen and not seen in my garden. If I don't take control, the juniper and wintercreeper will grow to block the sculpture whose location it seems I carefully planned.

Two years after we moved here, we pulled out a cedar tree that a hard winter's fickle frosts killed. Days after the removal, my parents delivered a sculpture they removed from the yard they left when they downsized. The sculpture's branches of curving and columnar steel balance a stylized orb. Thanks to the death of the arborvitae, we had a large hole in our dooryard where the seven-foot mass of crafted metal fit. How often happenstance looks like design. Why should passersby miss my magnificent, perfectly placed sculpture because of an unruly wintercreeper? I am the one with the shears.

Human ingenuity gave us power tools and hedge trimmers, herbicides, pesticides, irrigation canals, and plasticized landscaping fabric. Human ingenuity gave us the ability to enforce *The Great Chain of Being*. I deserve to revel in the spoils.

I dwell on the violence of our imaginations and actions sometimes. But the Bible I was raised reading, the one I am also teaching my daughter, claims a fundamental rule: "Love the Lord your God with all your heart, with all your soul, and with all your mind," Jesus is said to have said. "This is the greatest and the most important commandment. The second most important commandment is like it: 'Love your neighbor as you love yourself.'" This is the Word according to Matthew, Mark, and Luke. And Jesus took his cues from various passages of the ancient text, including the sixth chapter of Deuteronomy, a portion known as *Devarim*, or "words," in the Torah. "The whole Law of Moses and the teachings of the prophets depend on these two commandments," said Jesus. Love God with every part of your being, Jesus said. And love your neighbor as if your neighbor's being was the same as your own. It seems clear.

But who is my neighbor? Bob and Pam next door, yes. Also the neighbors with the trampoline, who have always been kind to my family. But what of the neighbor on the other side, who has never spoken to me? What of the neighbor who reported our compost bin to the HOA? What of the man with the DON'T TREAD ON ME flag whose street Ray and I stopped using as a shortcut to my parents' after he glared when we walked by his house? Spin the globe. Pick a place— what of the people there? What about the rabbits, or the wandering house cat who threatens them when he stalks in our yard? Birds and beetles? All birds? All beetles? Even those classified as invasive? What about the prairie dogs who stand outside their burrows and chirp to each other and, when I bike past their carefully constructed colonies, seem also to chirp to me? I love the prairie dogs unreservedly. But I had a student who called prairie dogs "sage rats." The boys in his family made a game of who could shoot the most in an hour. Prairie dog burrows pose a mortal danger to the family's cattle. The cows sometimes break legs in the rodents' incessant holes. The love part, that's what I want to teach Callie. I want the love part to be the most important thing. But the rule wouldn't bear repeating so often if it was easy to follow.

Let me not blame America's homogeneously cultured lawns and precisely trimmed hedges on seventeenth-century landscape architects alone. That would be unfair and untrue. In fact, a great deal of the blame goes to World War I and World War II.

Out of those wars came unparalleled tools: shelf-stable rations, cavity magnetrons, nerve gas, the atomic bomb, sound navigation and ranging (sonar), and more. From these, we developed canned foods, microwaves, microwavable dinners, and various preservatives for steady, abundant, unblemished meals. We developed neonicotinoids and other pesticides and insecticides and fungicides and herbicides to kill mites and weevils and beetles and mealybugs and mildew and mold and plants we don't want in our gardens. Like plastic, nylon, the computer, the space race, Teflon pots, the swivel desk chair, the jeep, and sonograms that gave rise to explosive gender reveals, the ability to keep lush, green, bug-free, weed-free yards has everything to do with the need to convert wartime technologies into daily noncombatant use.

But we are still at war. We combat anything and everything that does not receive our immediate favor. If you are a plant or a creature who reproduces without our sanction—say a lawn grub or a dandelion or an aphid or a slug—you might as well be the devil herself.

America applies one billion pounds of pesticides to fields and forests and rangelands and parks and lawns every year. We call these "conventional agricultural methods," as if this way must be the norm. A 2016 United States Geological Survey (USGS) study suggests that Northern Colorado saw the application of more than sixty-four pounds per square mile of the pesticide atrazine alone. "The use of pesticides has helped to make the United States the largest producer of food in the world," the USGS reported, "but has also been accompanied by concerns about their potential adverse effects on the environment and human health."

Neonicotinoids, a category of insecticide, are often applied to seeds and seedlings planted in household yards. They provide protection to plants for up to ten weeks. This biological agent, though, is indiscriminate. The nerve-corrupting influence of the chemicals the industry calls "neonics" control crop-compromising pests, but they also hinder the life cycles of the bees, butterflies, and small birds who visit fields and gardens. Hoverflies land for a taste of sugar-rich honeydew trails left behind by poisoned mealybugs and end up paralyzed, vomiting, dead. Honeybees pack pollen into their sidesaddle pouches and bring home nerve agents that decimate millions of hives. According to a 2019 article in *National Geographic*, America has created a landscape that is "48 times more toxic" to honeybees and other insects than the landscape that existed thirty years before.

We spray leaves, treat seeds, drench soil, inject trees, and anoint our dogs and cats to control fleas. When bugs fly down to feed as they have always done, or when centipedes wiggle through loosening earth, and when songbirds or tree frogs or lizards catch and swallow what they once relied on to keep themselves and their offspring alive, they get caught in a cycle of violence that spares no one. Pollinators come to America's gardens seeking sustenance and refuge. They fly away poisoned. Beneficial invertebrates like earthworms and sow bugs die on contact with even mildly treated soil. Birds who eat treated seeds, and birds who eat insects that encounter treated plants and soil, risk emaciation, disrupted reproduction, disorientation, and muddled migration patterns. This puts them at risk of slow or immediate death.

A report from the American Bird Conservancy reveals that "neonics in products sold for residential use on ornamental plants are as much as 30 times what's allowed in the agricultural sector." I might have bought plants treated with neonics at a big-box store's nursery. They are common but rarely labeled. I could be part of the problem and not know.

The umbrella term for pesticides and fungicides and herbicides is biocide. As with all doctrines, the doctrine of continuous combat with nature is often broadly and violently applied.

If it's not war exactly, it is a kind of combat nonetheless. If it is not hell down here, exactly, neither is it heaven.

Look at us on The Great Chain of Being. Clearly not on the bottom, but still not at the top. What with too little or too much water, crop-corrupting blights, harvest-ravaging pests, and beasts who care not a whit about us and our needs, even in my suburban backyard things are wild in this world. Why not use the rites and rituals and tools at our disposal to exercise control over the small things I can command? But also: Love your neighbor, I remind myself.

Their names sound like music: *Bombus nevadensis* (the fuzzy black-and-yellow Nevada bumblebee). Purple penstemon, silverleaf phacelia. Like in the movies, when a character leans in to hear a lover whisper in a different language: *Dalea purpurea* (purple prairie clover). *Populus tremuloides* (quaking aspen). *Solidago. Delphinium nuttallianum. Haemorhous mexicanus* (the three-quarter-ounce red-breasted, red-faced house finch who frequents our feeders). If you come to my garden, you'll see what I mean. Say them with me: Rocky Mountain bee plant (purple crazy-headed flowers on bright green, waist-high stalks). Little bluestem and sideoats grama (native prairie grasses). Pine siskins (little brown birds). Painted ladies (brown, orange, and white-spotted butterflies who feed on our hollyhocks, allium, echinacea, and hyssop). I do not tire of repeating the names of the many lives I am learning to love.

Maybe that's what drove Adam, what drove those naturalists who trekked westward after the Corps of Discovery: Douglas and Townsend and Nuttall. To differentiate each being in a field of chaotic color. Not so much the thrill of ownership, dominion. Simply the chance to recognize, to definitively recognize, another living being. Maybe Diego de Valadés delighted in engraving all those birds and fish and animals in an organized fashion the eye could understand.

I like to think I can separate the careful attention of naming, and the desire to care for and maintain, from the impulse to command and control. I like to think something I might call patient and abiding love keeps me working in the garden, watering with care, learning the names of the parts of creation I encounter. It feels different to walk into a room where I know everyone's name. To know the names of my neighbors is to know something of respect and affection and, sometimes, yes, love.

"Don't hurt that tree. Trees are people too." This is what it means to truly recognize another living being. What it means to love the greater-than-human world.

But lessons repeated all around me say human ingenuity—particularly, my own nation's ingenuity—makes it easier to exercise our supposedly God-granted dominion over the other living beings of the Earth. Fish are dumb and bugs are pests and weeds are weeds and the limbs of trees are there for us to break whenever and however we want. We are the ones in control.

We have the tools and the language we've put into God's mouth on our side. Listen to me name our helpers: indaziflam, imazalil, dicamba. These biocides that grant us the power of gods and of angels. Whether I use them in my garden or not, they surround me like the carbon dioxide–and methane-saturated air. Bifenazate, clothianidin, metribuzin, spinosyn. These human innovations that assure our command and control in the fields as surely as compacts and canals control water. Glyphosate, thiamethoxam, malathion, picloram. Heard with the right ear, these must sound like music as well.

Sunflower with a long-horned sunflower bee

Origin Story

Outside my window is the beginning
of half my poems. The others start
outside my door. In each case the window
is my body. I am always on
the other side of the door. All summer
every place around me caught fire.
The flames' orange haze spilled into my blood.

Bindweed, Convolvulus arvensis

The morning after the 2016 presidential election, Ray pulled our blinds closed. That November, our sunflowers were still bright. Still vibrant: the purple and burnt-orange chrysanthemums, the sunset and blue blazes hyssop, all the lovely late bloomers. Before the election, I opened my blinds every morning to look out at my garden all day.

But we can't know what passersby might think if they caught a glimpse of our home's interior. We hang carvings from the African diaspora and masks that my parents, my sister, Ray, and I collected from Ghana, Nigeria, South Africa, Zimbabwe, the Dominican Republic, Puerto Rico, and Afro-Mexican communities in Oaxaca, Mexico. Portraits on our walls, many by professional artists, show a satisfied, comfortable Black family.

On the morning after the election, white male CSU students surrounded a young Black man walking to one of Ray's classes. They aimed fists at Ray's student's face and jeered, "Obama can't protect you. Our guy's in now." And well before the election, back in 2014, a group of CSU students filed a complaint against bouncers who violently denied their entrance to local businesses. The only difference between the students tossed out and those allowed into the bars appeared to be that the students allowed inside were white. It wasn't until March 2017 that a man named Joseph Giaquinto smashed windows at the Islamic Center of Fort Collins and threw rocks, bricks, and a Bible inside. And it wasn't until August 2017 that the only Black resident in a CSU dorm woke to find a noose taped over his hall's entrance. And still later, in May 2018, a white mother touring our university called campus police to search and detain two teenaged Native American brothers who joined the tour group late. "They don't belong," she said. But by the morning after the election, as a professor in the Ethnic Studies Department, and as a Black man in our community, our campus, our country, Ray already knew too many terrible, terrifying stories about what could happen if a white passerby looked in our window and decided we didn't belong.

We didn't close the blinds that November because Ray was afraid we'd be robbed. He was afraid we'd be killed.

The Sunday after the 2016 election, my mother wrote a prayer card seeking protection for many people: Black people, Latinx people, LGBTQ people, women and children, refugees and other immigrants, people of non-Christian faiths—especially Muslims—disabled people, and people with health concerns. When the pastor finished reading Mom's long list of people to hold up in prayer, he added a sentence for context: "Let us pray for those who are on the outside of our society looking in."

I looked around the sanctuary, hoping for evidence that someone else seemed shocked by his language. Besides my family, no one seemed troubled.

After the service, Callie joined the other children in the youth choir. The week before, the choir director pressed manicured hands on her updo, checking for stray dyed-blond strands. She told me the children looked up to Callie, who was six at the time. They followed her lead because she could read. "It's nice to have her in the group," the choir director said.

There were only about ten kids under the age of thirteen at the church. As with many mainline Protestant churches, the congregants were mostly over sixty. We first visited at the recommendation of friends of my parents and stuck around because Callie connected quickly with the six-and-under crowd. I liked being part of the pastor's efforts to welcome new generations into fellowship. On the Sunday after the election, my brown-eyed daughter and the pastor's blue-eyed girl sat side by side in the children's choir, working on lines for a new song.

I grew up singing in church choirs, attending Sunday school and youth group. I was confirmed in the church, and married in the church, and, back in Oakland, Ray and I belonged to a congregation that loved Callie since she was only a bump in my womb. I turn to the church for just such a sense of community. The song the children practiced that November Sunday was new to Callie and her friend, but not to me.

Sometimes we get early snows that leave trees denuded and flowerheads dead, but that November week, daytime temperatures reached the high 70s. Red and pink cosmos, bright as neon signs, swayed waist high in our lawn's central plot. Amethyst four o'clocks (*Mirabilis multiflora*, which means "wonderful multiflowered plant") stayed open late into cool evenings. And bindweed, which starts in spring and continues until snow flies, grew strong.

Bindweed's taxonomical name, *Convolvulus arvensis*, means "to entwine the field." I first noticed the flower opening on vines around the yard in early June 2014. I thought it was a type of morning glory. The two plants are related. The bindweed's flared trumpet flowers measured an inch in diameter. They looked delicate. Some pure white, some dusty pink. A few displayed an ombré gradation through white to pink before settling on a purple-blue shade near the flower's nectar-filled well. I let the plant grow and grow and grow until I noticed the way the vines endangered other plants in the yard. Bindweed's long tendrils spread aboveground, away from the rootstalk, in many directions. Grass fronds, snapdragon, plant stakes, blue flax, the limbs of the dwarf mugo pine, anything the bindweed touches, it wraps and binds and climbs.

The same way it stretches aboveground, bindweed sends its roots twenty feet into the soil, outcompeting other plants for hard-to-find water sources in dry land. The plant is rhizomatic, with an ever-growing horizontal underground stem that can move laterally as much as ten feet in a season. These rhizomes sprout new feelers and roots, seemingly at random. Versatile growth patterns allow bindweed to find the most advantageous spaces to proliferate. The word for this kind of erratic and copious root system is *adventitious*. Adventitious roots allow bindweed to propagate at any point or points along the way. Cut a leaf, the roots, a tendril, and new roots will emerge from the broken parts. Tilling and careless pulling only spread bindweed more vigorously. Bindweed can survive under- and aboveground. In dry conditions or wet. With or without direct sunlight. In too-poor conditions, bindweed bides its time, lying dormant for as long as forty years. Bindweed's roots web the top layer of soil with widespread clumps of white filaments that suppress the growth of other species.

. . .

The pastor's sermon the Sunday after the 2016 election prioritized getting along with people who might have voted differently a few days before. He stressed the importance of maintaining a comfortable civility, suggested we practice the art of turning the other cheek when someone slaps us. If your assailant asks for your coat, the pastor quoted from scripture, give him also the shirt from your back.

Attempting to placate a congregation that did not want to deal with the trauma of pulling apart, the pastor quoted a sermon by Dr. Martin Luther King Jr. He did so without apparent cognizance of the fact that two people in the church had been young Black adults when a living King delivered the sermon himself. Except for two graduate students from Borneo, my parents and I were the only nonwhite people in the sanctuary. Had the pastor taken us into account, perhaps he would not have asked us to give away also our shirts when stripped of the liberties we've fought our whole lives to attain and defend. He spoke to the white people in the room using King's language, but he corrupted the core principles of the resistance movement out of which King's words arose.

The Black Americans who employed nonviolent resistance in the long battle for civil rights publicly refused to accept derogatory and damaging physical, legal, and cultural attitudes and behavior. The nonviolent resistance movement pushed the antisegregation struggle into America's living rooms via the television screen. White people, who benefited from a system designed to keep them ignorant of the toll of racism on Black lives, had to confront their complicity in segregation's violence.

When white policemen aimed fire hoses at Black marchers, flaying protesters' skin; when the parents of five hundred white students refused to let their children enter their New Orleans elementary school because one Black six-year-old girl enrolled in the first grade;

when white teachers at that school ostracized Barbara Henry, the sole white woman who taught Ruby Bridges; when student and teacher sat alone in their isolated classroom while mobs of segregationists booed and chanted outside; when white men placed explosives under the steps of 16th Street Baptist Church, a meeting place for the Black community in Birmingham, Alabama; when the explosion killed three fourteen-year-old girls and an eleven-year-old girl, nearly blinding a fifth child; when white men firebombed Black leaders' houses or burned crosses in front of the homes of Black activists and their families—in these circumstances and so many more, defying racism required courage and direct confrontation. The nonviolent resistance movement actively called for white America to look at this country's systemic violence toward Black people. In one brutal 1965 confrontation, now known as "Bloody Sunday," state troopers with tear gas, clubs, and even rubber tubing studded with barbed wire, attacked Black voting rights activists as they tried to march from Selma to Montgomery, Alabama. Frightened but resolute Black men and women immediately regrouped and planned to undertake the march again. Dr. King told that crowd, "If you are beaten tomorrow, you must turn the other cheek."

The nonviolent resistance movement demonstrated white America's culpability in the slapping of all those cheeks.

This congregation had a tradition of congratulating the pastor on a well-delivered sermon. A manifestation of the convivial community I desired. After the service, I stood in line with the rest of the church.

Tears in my eyes and my throat, I told the pastor, "The language you used during the prayers of the people was hurtful and dangerous."

He looked wounded and surprised.

"All those people you listed," I continued, "we are not 'on the outside of society looking in.' We are part of this society! We are at the very center of what America has been built upon. But the rhetoric you used during your addition to that prayer is the rhetoric of exclusion."

When I find a patch of bindweed growing around the blue flax in the garden, I can't just tear the strangler up at the root. Doing that will leave tendrils twined around the blue flax stems, choking the plant I want to support. Sometimes, if I try to remove the bindweed too quickly, I pull up the blue flax too. Since bindweed can regenerate so easily from broken pieces, I am careful when I weed it out. I start from the top, where the bindweed is newest, and carefully unwind the tendril from the plant it chokes. Sometimes several tendrils wrap around a single stalk, pulling at the blue flax from different directions. I separate all the pieces consciously and carefully, so as not to damage the blue flax or leave pieces of bindweed on the soil to resprout.

Unable to hear my conversation with the pastor, the line of white congregants stalled behind me. Out in the fellowship hall, the buzz was loud and indifferent to us. The pastor no longer shook my hand in the normal, absent-minded manner of a reception line. We stood still, hands clasped together, as if we might begin to spar or to hug.

Disgusted that I might look hungry for this white man's affirmation, I finished what I wanted to say with force, "You mentioned *women* in that! Women make up more than half the population. We are not outsiders looking in!"

He apologized. He insisted I should always tell him if something he said or did was hurtful. I looked into the pastor's ungrasping blue eyes and felt like a fool. These must have seemed generous gestures to him, but he took no agency. He made *me* responsible for calling his attention to the hurt he caused.

In 2013, Public Religion Research Institute (PRRI) conducted a study revealing that white Americans build networks of close friends and family that, on average, consist of 91 percent other white people. The average white American's self-reported social network is only 1 percent Black. It's also only 1 percent Hispanic, 1 percent Asian, and 1 percent mixed race. The rest of the nonwhite people in the average white American's social network represent some "other race" (1 percent) or a group PRRI defined as "Don't know/Refused" (3 percent). The figures presented by PRRI don't add up to 100 percent. When it comes to race in America, not everything adds up. The PRRI research reveals systemically constructed failures of the integration of white America's social networks.

In many cases, several people within a white social network count the same Black friend. Though I might have counted nine white friends at that church, when asked if they had any friends who identified as Black, all nine of them may count only me. Like the blue flax choked and torn by several bindweed tendrils, I could be surrounded by people who seem at first to mean no harm.

Those PRRI statistics reveal only averages. Averages that *inflate* the number of nonwhite friends and family claimed by most white Americans. "Fully three-quarters (75%) of white Americans report that the network of people with whom they discuss important matters is entirely white, with no minority presence," write the study's authors. Fill a room with one hundred white Americans and seventy-five of them will have no nonwhite friends or family. None at all.

Because the average white American knows so few nonwhite people, they also know little about the traumas that systems of social and racial segregation and oppression enforce. Rather than working to widen his own circle of experience, the white pastor asked me to translate my experience into terms he could more easily digest. That kind of laziness does substantial damage.

In the months after the 2016 presidential election, I often found myself in the company of people, almost always white, who said, "This is all so surprising. This isn't who America is!" Every time, I found a way to move my body as quickly as possible out of the reach of their mouths. I heard such statements again when a May 2017 Montana special election proved a white man could win a congressional seat despite pleading guilty to assault charges. Because many voters had cast ballots before they knew about the winner's violent tendencies, some people said the outcome of that election was skewed. Others professed shock at the level of violence American voters condone. "I didn't think this could actually happen!" people said. But I was not shocked. For quite some time—since the beginning, really—Black Americans have pointed out that "this" is actually happening. Expressing shock at the problem at this point felt like too little too late.

Callie liked the huge playground in that church. On warm Sundays after youth choir, she and the pastor's five-year-old daughter took turns on the slide and the swing set. They tumbled and cartwheeled through the grass on the wide lawn, even in their Sunday best. If I warned them to be careful of staining their dresses, they sat together making bracelets from clover flowers they'd picked out of the green expanse. The girls spoke the shared language of childhood: giggling, whispering, playing pretend.

Hand in hand, the girls scampered toward the parking lot most weeks when our time at church was over. If I unlocked the door before they arrived, they'd both climb into my car, the pastor's daughter huddling on the floor below Callie's car seat like a small pet who hoped to be immovable. Knees to her nose, forehead resting on her pale wrists and fingers, her mop of wispy hair fanned out on the car's floor. Sunday after Sunday, and once after a church dinner at the pastor's house, the pastor had to physically remove his daughter from my car. "I want to go home with them," she told him. "I don't want them to leave."

While I stood in the narthex talking with the pastor the Sunday after the election, my daughter played with his daughter, learning the basic tenets of the church by coloring in a black-and-white picture of a dove and a flame and crossed planks cut from a living tree. I walked into the youth room already announcing that we wouldn't be going to the playground. We had to leave right away. The girls ducked into a hiding spot and stayed as quiet and still as they could, their small limbs and smiles pulled close in the dark.

Before I reached the girls' hiding spot, the Sunday school teacher stopped me. She held up the worksheet with the denomination's icon and asked what I saw. "The flames of the spirit," I said. "And the dove." I pointed out these elements. Both suggested the promise of new and sustaining life. "In the Cross, I see the open arms of Christ."

The teacher looked surprised. She said she didn't think most people thought about what the image contained. Moments later, the pastor's wife came into the room. As if to confirm her point, the Sunday school teacher asked her the same question. "That's the symbol for the Presbyterian Church," said the pastor's wife. She seemed caught off guard by the question. Her cheeks paled then reddened as she spoke. "I don't know what it means."

Her answer surprised me, but it shouldn't have. While I need to read meaning into everything I see, she didn't need to think twice about the signs that identified her community.

After the 2016 election, I worried about people's intentions almost all the time. I worried people I liked would say or do things that, though innocuous to them, caused me great pain. I worried that even well-meaning people would cast me aside. I worried that speaking about my worries would alienate me further from my community of white friends and neighbors. Energy spent correcting an "I didn't think this was who America was!" here and a "This is so surprising!" there drained me. More and more often, I stayed in my house.

Not long after the January 2017 presidential inauguration, we moved down to the basement while we renovated our master bathroom and insulated the house's ceilings, floors, and walls. Our house was torn apart. Like plants that pull into their roots during the harshest months, we hunkered underground. This felt remarkably right.

Callie slept in the basement room the sellers' real estate agent had advertised as a craft room. We call it the Dot Room, because when we renovated the lower level soon after moving into the house, I covered the room's original white color scheme with chalky blue wall paint, adding a series of contrasting circles in gray and cream and black chalkboard paint. To trace the outlines of these circles, I used avocado-green lazy Susans my mother had held on to since the 1970s. She gave them to me when they downsized. The lazy Susans made perfect stencils. I still haven't thrown them away. I come from people who don't throw things away.

We use the Dot Room as an overflow guest room. It has no

windows. Concrete sits just behind the drywall. The room stays cool and dark. It makes the perfect place to retreat. The Iowa house I lived in as a teenager, after we left Southern California, had a similar room. That sellers' agent advertised it as a 'fraidy room. My parents kept wine and a few boxes there (perhaps storing the old lazy Susans), but mostly we found ourselves in the 'Fraidy Room only if tornado sirens sounded.

What am I doing?

I'm trying to describe an experience of being a Black woman in America. What I'm describing are panic rooms.

There's a line in my essay collection *Guidebook to Relative Strangers* that caused concern when that book headed to press. In the line, I address the fact that my grandmother's father was, as one direct descendant described him, "damned near white." A few of my white friends, trusted early readers, wondered why his near-whiteness didn't get him a pass from the violence directed at darker-skinned people. Shouldn't his near-white privilege have kept my great-grandfather safe?

That's not how it works in America.

In America, Black isn't a skin tone. It's a condition.

That great-grandfather had a successful plumbing and sheet metal business. He made enough money to hire a younger cousin as his apprentice, and together they completed work for some of the grandest homes in town. But he woke early one morning to open his shop and found his cousin's body thrown over a worktable, along with a menacing note. My great-grandfather was a Black man in America. No matter who his father was or who his father's father was or how close he toed the line to whiteness or how much he believed in his right to advance beyond the menial stations reserved for Negroes in this country, white men in that town made sure my great-grandfather knew he wasn't going to be allowed out of the limits imposed on his Black body. Except through death.

On April 29, 2017, while my family still slept in the basement, police killed another child.

Fifteen years old, his name was Jordan Edwards.

Jordan left a party with his brothers, but rather than spend the rest of the night celebrating the successful end of another year of high school, Kevon and Vidal Edwards witnessed the last moments of their little brother's life.

In the months after the shooting, people pointed out that Jordan was an honor-roll student, as if academic success should have granted him a pass from the condition of Black lives in America. Officer Roy Oliver, who murdered the child, claimed Jordan's car advanced aggressively toward his fellow police officers. Video footage disproves this assertion. I'm tired of these conversations. I'm tired of discussing whether or not police should find it acceptable to kill a fifteen-year-old Black boy while he sits in a moving car.

Around the time Oliver shot Jordan Edwards, a white student from one of our local high schools contacted me. This was a sweet girl from a family I liked. She needed to interview a professional woman whose work she admired, and I happily obliged.

At the end of her lighthearted interview, the girl asked a final question: "Why do Black lives matter?"

Her tone made it clear that she earnestly wanted to make sense of a movement she could not understand. There were only twelve Black students in her entire eighteen-hundred-student high school. She couldn't understand *why* police disproportionately kill Black people. More crucially, she couldn't understand *that* police disproportionately kill Black people. She had very little exposure to American history as it relates to Black and white relations, and she hadn't heard many stories about Black lives in America. So she couldn't understand the imperatives that drive the Black Lives Matter movement.

I gave this child—who was the same age as Jordan Edwards—a sound bite and a few statistics about the number of Black people and people of color law enforcement officers kill before these Black and Brown people have a chance to access the due process of law. She could use this information in her report. Even a slight comprehension might make a difference in her very white class in our very white town.

I didn't curse her question, though I wanted to then and have wanted to many times since.

Do you think your life matters? I wanted to say. *Well, that's how I feel about mine.*

I think Jordan Edwards's life matters. I am angry he is dead.

In her book *Kingdom of the Blind*, the author Louise Penny wrote, "The real danger in a garden came from bindweed. That moved underground, then surfaced and took hold. Strangling plant after healthy plant. Killing them all, slowly. And for no apparent reason."

Complaints about bindweed have shown up in texts since the ancient Greeks. One sixteenth-century English botanist wrote, "Bindweeds are not fit for medicine but are unprofitable weeds, and hurtful unto each thing that groweth next unto them." First recorded in North America in 1739, bindweed probably arrived in Virginia with a slaving ship's cargo. Archeologists discovered it in adobe bricks used in the Spanish settlement of California. By the

1890s, thanks to the western expansion driven first by the Corps of Discovery and then by the railroad, bindweed spread throughout North America. Though it sprouts nearly everywhere in our yard, the weed is classified by the Colorado Department of Agriculture as a List C species that should be "either eradicated, contained, or suppressed."

I've come to understand that I'll struggle with bindweed, one way or another, until I give up. Or I die.

Dear Brenda:

I have started writing again after the great silencing that was the election and the installation of that person and all those who stand behind him and his supremacist notions. Now I am carrying a very small, very cheap, very inconsequential notebook. And sometimes I write things in it. It feels like something. And it also feels like nothing. Like all those calls I keep making to my congressman and senators.

Here is a photo of the little gifts you sent Callie. They are leaning against a start from one of my African violets. All of the violets have done so well in the living room that they outgrew their pots. I've had to give them new homes. They're struggling now. But they are hanging on.

My whole life feels like a metaphor these days.

All love,
Camille

I write it all down. In notes, poems, essays, in messages to friends. It's the role of the artist to observe and record what happens in the world and to whom. What we do not see, we cannot correct. What we do not acknowledge, we cannot repair. One of the most powerful tools of oppression is the insistence that certain lives are of little consequence. That some people's words are inconsequential. That what they grow and raise and build and love are inconsequential. That what matters to them need not matter. Such categorical dismissal is not easy to achieve. Day by day and year by year, such cruel power takes a long time to root down. And even longer to eradicate.

To protect himself, his wife, his children, my great-grandfather moved miles away from the town where men killed his cousin. He built a new home for his family—what he hoped would be a safe house—with the only indoor plumbing in town.

He opened a sheet metal shop on the border that divided a dry state from a wet one. On the dry side of the shop, he built bathtubs. On the wet side of the shop, he built stills.

I'm pretty sure I know why I love this last part of the story so much. I'm pretty sure it has something to do with a Black man bending the law of the land to his will.

The service I sat through the Sunday morning after the 2016 election worked such a violence on my spirit I have not brought myself to go back to that church. I was struck too hard that Sunday to take the gamble. I couldn't trust that pastor anymore. Maybe he understood this when he didn't see my family the next Sunday or any Sunday after.

For more than a month after the election, Callie cried on Sundays. She wanted her friend. *Couldn't we set up a playdate?* Callie asked so often I finally sent an email. It took the pastor months to respond. No one from that congregation ever called to ask about our absence. I told Callie there were other churches where we might build community. But I couldn't get her excited about attending any of them. She already knew a church where she thought she belonged.

The rhetoric of exclusion that emboldened a presidential campaign, a whole administration, and, in many ways, a nation, caused immeasurable pain. For reasons she may never understand, this rhetoric robbed the pastor's daughter of a friend.

Those adventitious roots, the rhizomatic structure, the initially innocuous—even lovely—appearance of its flowers, the flowers' copious seeds, how the plant bides its time: these keep bindweed growing even in a well-tended yard.

In the late fall of 2016, I watched a woman a few blocks away prepare her house for the market. The house was a rental property. No one had tended the broad brown mulch patches in the front yard for several years. When I passed the house on my way to drop Callie off and pick her up from school, I watched this woman weed and weed and weed. Trowel in hand, on her knees, she dug up thistle and bindweed, piling the evidence of her labor as much as eighteen inches high.

Through the winter, the yard rested. In summer the landscaping looked cared for and loved. But, growing up and around struggling sage bushes and pulling down rudbeckia stalks, the bindweed came back in less than a year.

I can recognize bindweed from a few arrowhead-shaped leaves on new tendrils. If I keep on it, I can reduce the damage it wrecks. I can prevent the emergence of new plants and limit the growth of old ones. But I have to watch for bindweed almost every day. I could spend hours each week, most weeks of the year, worrying about bindweed.

I visited Callie's elementary school sometime close to Memorial

Day 2017, to help the children in her class with their reading lessons.

It felt like the first day of spring. On the schoolyard, kids from a class ahead of Callie's played at recess. The girls wore sundresses. No one wore socks.

I noticed the grass had turned green.

Four blond girls played tetherball. A Black girl stood off to the side. I thought she said, "Can I play the next game?"

I thought I heard one of the blondes answer, "Sure. Tomorrow."

On the steps to the school grounds, someone had carefully chalked the words

THE WORLD IS CHANGING

I wondered whether to read this as a threat or a promise.

Callie with a quaking aspen leaf

Around the time in late May 2017 when we moved out of the basement, our handyman helped me by hauling several cubic feet of rock and a bucket of bindweed from the yard. After that, I prepared some soil to receive pole beans a friend had gifted me the previous summer. Beans from a line of seed passed on by survivors since the 1838 Trail of Tears. Together, we made a space in the garden for something that would look, by autumn, like edible hope.

I am getting ahead of myself. Working the land, I lose track of where I am in time. What happens today is fed by what I did yesterday. What I reaped that fall recollected decisions made by the likes of Dr. John Wyche, the man who first sent heirloom Cherokee seeds to whomever showed interest and paid postage in a decade I was nearly too small to remember and that Callie calls the olden days.

If I started from the start, where would I go? Black-eyed peas, a staple food in West Africa, made the journey with enslaved people from that continent to the American South. In their book, *In the Shadow of Slavery: Africa's Botanical Legacy in the Atlantic World*, Judith Carney and Richard Nicholas Rosomoff write that these same people used the stimulating kola nut to manage fetid water they had to drink on slavers' ships. Later, that nut became a key ingredient in Coca-Cola. When I speak about garden-variety crops in America, I always point toward simultaneous legacies of trauma and triumph.

Watermelon, sorghum, sesame seed, rice. The millet I throw down for gray-feathered, ground-feeding dark-eyed juncos. None of these would be what they are in America but for centuries of human trafficking known as the slave trade. Some ancestor kept okra seeds in her hair through the long trial of the Middle Passage and onto, *into*, American soil. Someone secured raw peanuts in an unsearched scrap of cloth near her body. Peanuts, like pole beans, like black-eyed peas, are both food and seed. I can eat them for power today or plant them for abundance tomorrow.

People who came long before me carried the source of a new kind of flourishing through desolation I cannot fully comprehend. If I say my garden's story starts with the planting of a seed, to which seed am I referring?

I remember the first garden I planted as a married woman. It wasn't much to speak of, neither the garden nor the house in front of which I sowed it. But the garden helped me feel rooted in the place where Ray and I began our new life. I planted marigolds and nasturtium. I put in zucchini for its riot of bright blossoms. I kept an artichoke whose purple thistle flower delighted me, though an army of ants quickly moved the heart beyond the possibility of human consumption.

Even if I had managed a harvest during our seven months in that house, I shouldn't have trusted food produced from that dirt. Fumes from the nearby freeway drifted over us night and day. Paint flakes flew from the 1923 duplex's exterior walls. Soil tests in the area have revealed lead levels hundreds of parts per million above safe limits, and I didn't build raised beds. Still, I wanted to witness a plan come into fruition. I planted two-dollar seed packets, four-dollar starts. I watered. I weeded. I watched. That Oakland yard boasted little but dirt before I started digging. For our brief time in that house, we walked out into a flowering.

I once shared a few hours with a Salvadoran poet who walked across the desert into the United States when he was nine years old. The landscapes he walked out of and across delivered incredible pain. Yet we talked about the importance of writers of color celebrating the living world. He recalled his grandmother's garden. *There was joy there*, he insisted. He wouldn't let his charge to document suffering stop him from recalling true pleasure.

As I dug in dirt contaminated with legacy pollutants outside that first home I shared with Ray, a woman from the neighborhood stopped to watch me. Her press and curl shined in the Oakland sun. She wanted to know why I bothered tending such a yard.

A local nursery's discounted marigolds slumped in their black plastic pots near my knees.

I remember feeling so angry.

Our block, our rented house, deserved such care. "I know it might take a lot of work," I told her, "but I want to grow something beautiful."

The Earth's produce can provide a lasting record of access, autonomy, and power. Eighty miles southwest of Monticello, Thomas Jefferson kept an estate called Poplar Forest that he used as a retreat. Caches of food found in storage pits on the estate reveal the epidemic of deprivation endemic to the enslaved people who lived on the property. The caches also document the survival strategies of people insistent on nourishing themselves.

Archeological studies suggest that the people at Poplar Forest grew corn in their gardens. Remains reveal the culinary use of wheat, oats, rye, sumac, blackberry, purslane, pigweed, poppies, and more. The people Jefferson bound to this property probably grew sunflowers, mint, sweet potatoes, and violets. They might have grown the violets and sunflowers as ornamentals but, just as likely, they used them for food. A quarter of a cup of roasted sunflower seeds contains up to fourteen grams of healthy fats, five grams of protein, 37 percent of a daily allowance of vitamin E, and 17 percent of a daily allowance of folate, which promotes health in pregnant women and babies. Violets can serve as a replacement for okra and collard greens, plants rich in carbohydrates, protein, calcium, iron, and vitamins C and B6. Even the ornamental plants around the quarters served as edible provisions for the people who tended the land. I like to think that, as much as they depended on them for physical sustenance, the people who lived in the quarters at Poplar Forest liked looking at these plants as well.

The people raised chickens, whose eggs they could sell. They may have also sold other produce from their gardens. For a time, they grew another cash crop, but Jefferson made sure his son-in-law "put an end to the cultivation of tobacco" by the people he claimed as property. "There is no other way of drawing a line between what is theirs & mine," Jefferson admitted in a letter. He forbade these men and women from growing for personal use the same crops they cultivated in his fields. I have histories like this in mind when I insist on growing what I please in the soil that surrounds me. There is power to be generated from cultivating whatever might sustain me in whatever way I wish.

I grow sunflowers and sweet potatoes in my garden. I plant what plants I desire, and I harvest or not as I choose. I grow mint and tolerate the purslane people these days tend to weed out. I've considered tattooing a line of Lucille Clifton's poem "mulberry fields" on my own flesh: "bloom how you must i say." Sometimes unmastered growth reveals our dearest needs. I grow poppies and let wild violets flourish—for through their flowering, time progresses.

It's longer than Callie's whole life since our Oakland neighbor questioned me, but I still regularly encounter incredulity when I talk about coaxing beauty out of the legacy pollutants that haunt us. Once, after I delivered a public reading from one of my books, a white woman in the audience asked how I could fancy myself an environmental writer when I write so much about African American history.

For a breath or two, I was speechless. I did not understand how I could write about history *without* accounting for the environment out of which history springs.

"The woman does not understand the fundamental importance of her personal or ancestral status as an immigrant," my dad told me, a few days after the encounter. Immigrants chose to come to this country, Dad said, "for a variety of reasons, including pursuing opportunities unavailable in their country of origin and escaping persecution or violence. That decision also provides the opportunity to decide how you do or do not interact with the environment and/or issues of social justice."

As we waited for his granddaughter to come downstairs, my father continued, "For those whose ancestors did not come to this country by choice, instead of new opportunities, they faced oppression." Most African Americans, he said, are descended from people who "were brought to this country as chattel, without consent, to 'tame' every aspect of the environment for the pleasure of others."

We heard Callie round the corner landing, hurrying to hug her granddad. "For us," Dad finished, "there is no separation between the environment and social justice."

Living in this body, I can't help but see the devastating implications of the erasures of certain histories.

The reason city planners so frequently ran freeways through the Black part of town and not others is because the lives and property of those who lived in that part of town were not valued. The pollution of that indifference persists in the very ground people walk on today. Just as I found it necessary to beautify that patch of dirt in front of the first house Ray and I shared, writing about the environment and discussing social justice are necessary political decisions.

When ants announced their interest in the artichoke in our Oakland yard, I let them enjoy its substance while I settled for appreciating its splendor. I did not depend on the artichoke for its nutritional value. If I want to sow an ethic of generosity and grace, sharing with the ants could be part of my goal. I refuse to take part in the segregation of the imagination that assigns greater value to some experiences than to others. If I cultivate a flourishing, I want its reach to be wide.

Near where I planted the Cherokee Trail of Tears pole beans, rhubarb greets me each spring. Scorned by many and by others fiercely loved, rhubarb is tricky. The broad green leaves are inedible, containing high levels of an irritant called oxalic acid. But rhubarb's fibrous and nutritious red stems contain many useful characteristics—with antioxidant, anti-inflammatory, antibacterial, as well as cancer-, diabetes-, and heart-disease-fighting properties. Those edible parts can taste bitter, though. Contemporary recipes add quite a bit of sugar to help the medicine go down, converting the vegetable into something used in simple syrups, cakes, and pies.

Who planted this rhubarb? Nothing else like it grew in our yard when we moved here. The plants' presence in the garden changes how I understand the people who lived here before us. What they cared about tending and how. I love this house, and this green-and-red leafy bush is part of why. The rhubarb comes back each year to remind me that, even under what appears to be dirt, I might find the root system of some kind of insistent thriving.

I never know how much I need to see that rhubarb unfurling until it begins to unfurl.

In June 2017, for the first time in our four years in the house, the plant burst into flower. The many-headed bracts looked like ten thousand snowflakes held firm on summer branches.

Garden advice suggested I should lop off the flowerheads to encourage the edible stalks to keep growing. The plants would go dormant sooner if I left those bold bids for pollination, and the rhubarb would be of no practical use. But, in their full blooming, the joy I gleaned in the garden erupted over every inch of my life. I let those enormous, lovely flowers be.

I opened our blinds so we could see.

Rhubarb gone to seed

In her mostly white town, an hour from Rocky Mountain National Park, a Black poet considers centuries of protests against racialized violence

Two miles into
the sky, the snow
builds a mountain
unto itself.

Some drifts can be
thirty feet high.
Picture a house.
Then bury it.

Plows come from both
ends of the road,
foot by foot, month
by month. This year

they didn't meet
in the middle
until mid-June.
Maybe I'm not

expressing this
well. Every year,
snow erases
the highest road.

We must start near
the bottom and
plow toward each
other again.

A cluster of flowers in the central plot

"I think that's a weed," Mom said. The late-spring sun warmed my exposed arms, but Mom wore a cable-knit sweater set in a shade of royal blue that complimented the lustrous deep melanin of her skin. She pointed to something that looked like a dusty Palm Sunday frond sticking out from the dirt near the dooryard's juniper and wintercreeper bushes.

"It is a weed," I told her. "It's milkweed. I've been hoping it would grow."

That swamp milkweed didn't flower the 2017 summer my mother first noticed it, but years later, in 2020, the pink-petaled plant thrived among the hearty survivors crowded into the dooryard. Hollyhocks; purple penstemon; white-petaled, yellow-eyed Shasta daisies; upright prairie coneflowers; purplish-pink echinacea; and big red fireballs of bee balm like something out of a sci-fi film about affable, non-Earth-dwelling sentient beings. Finally mature, this cluster of plants grew in differently from what I imagined when I dug up rock beds and threw down seed. It takes three years for plants in a garden to achieve their potential, still longer to fully establish themselves. Working in this yard, I've learned a great deal about biding time, about accepting that change does not happen overnight and often shows up differently than I expect.

On March 2, 2020, the first day they were available for sale, I ordered the two flats of plant starts I had thought about since I first heard of the Garden in a Box program in the spring of 2019. For less than $200, I bought enough native and nonaggressive naturalized selections to fill the prairie project on the south side of the house. The company opened orders early to gauge how many and which kinds of plants to propagate, but they didn't deliver my two flats, each with fifteen species, until late in the afternoon on June 16, when threats of frost had passed in Northern Colorado. For most of the thirteen months between getting the idea for the pollinator patch and actually planting my garden, preparing for our prairie project meant being willing to wait.

Then, a flurry of activity. By the evening of June 17, I finished putting everything in the ground.

The Garden in a Box's Colorado Oasis flat came with yellow columbine; blue grama grass; blue native harebell, a small purple bell-shaped flower that hangs from ten-inch wiry stems; false indigo (*Baptisia australis*), whose purple clusters of lupine-like flowers rise from shrubby green foliage; firecracker penstemon, with columns of twelve tubular scarlet flowers, perfect for hummingbird beaks; purple Rocky Mountain penstemon rising from crowns of two-foot stalks; prairie red coneflower; and switchgrass (*Panicum virgatum*), a native grass whose elegant three-foot panicles turn a deep burgundy in fall. Nothing grew taller than five inches when the flats arrived. But plants grow just as humans do. It amazes me how much can come from how small a start. Many afternoons in 2020, Ray and I watched the pirouettes of five-foot-tall Callie fill our living room, amazed he could once balance her entire body on his forearm while holding her whole small head in his palm. Profusions of red, purple, yellow, orange, green, and gold plants, as wide as four feet and as high as my knee, would fill our side yard someday soon.

I go overboard sometimes, crowding my gardens with too much. Planting the prairie project, though, I left thirty-six inches between each start's base. In that new plot, I wanted every plant to have sufficient space to spread and mature without compromising the growth of its neighbor. I placed tall and dense plants toward the back of our sight line, low plants near the front. I arranged plants based on when they flower, so we humans could enjoy a continuous display through the seasons, and so the pollinators I hoped to attract could more efficiently navigate from blossom to blossom throughout the year.

When we decided to remove the sod from the south lawn and start the prairie project, I let my neighbors Pam and Bob know. Not to ask permission, but because they would also have to watch the progress. The prairie project is within both of our views.

When they moved to the neighborhood a year after we bought our house, Bob and Pam inherited a row of fourteen rosebushes along the margin between our yards. More than once I've heard them sigh over the process of caring for the finicky red flowers, thorny stems, and red-green leaves. Susceptible to fungal infections from too much water on the leaves, liable to develop powdery mildew, rust, and other fungal diseases from poor airflow, and prone to parasitic insects who bore into the cane, roses ask a lot of the people who care for them. Supporting lasting beauty isn't easy. This is something anyone who has put time into such efforts knows. Despite the demands, Bob and Pam take excellent care of the bushes, cutting them back in fall to avoid cane break in the wind and snow. Clearing competing vegetation from around the rose's roots. Trimming and pruning to make room for new growth and ideal airflow. Even telling the rosebushes they look beautiful. I love a person who talks kindly to plants.

In their first years in that house, Pam gave me columbines and wild violets she pulled from around the roses' roots. I planted these in my own yard. Bob and Pam have descendants of our sunflowers growing in their front yard, away from the roses. We've built a friendship around flowers, complimenting one another and exchanging tips.

"Feel free to spread some of that mulch over this way," Bob said one April 2020 day when he saw me pulling overthick layers off our prairie project plot. When I planted the Garden in a Box starts in June, I distributed some of that excess mulch around their rosebushes, creating a visual tie between both our lots. Afterward, I sent Pam a text, listing what I planted.

She wrote back, excited. "I've been watching the progress with interest."

"There's more!" I replied. The Naturally Native box came with dwarf blue rabbitbrush; a low-growing land cover called pussytoes; little bluestem grass, a showy cluster that will grow to four feet; native gayfeather, whose wands of fuzzy, lavender flowers rise from a mint-green crown; native lavender bee balm; *Anaphalis margaritacea*—small white wooly balls growing around yellow disks—commonly known as pearly everlasting; little orange teacup-like prairie mallow; and wild blue flax. Plus, I put down milkweed seeds and poppy seeds and seeds from purple wine-cups, a pretty flower people sometimes call buffalo rose. To welcome digger bees and long-horned bees, native pollinators who favor many of the plants in the plot, I cleared more mulch and exposed patches of bare ground.

It astounds me that I managed to finish all of this in less than four hours, except that we'd done so much to prepare. I waited over a year for that day.

. . .

Planting the south plot, I kept thinking of my father. My careful, caring, patient father. Like the silvery-green Mojave sage that emerged at the edge of the front yard's central plot in 2017 and, with little tending from me, has grown ever more robust and fragrant since its first appearance, sometimes it seems I came to be who I am all on my own. I didn't plant the sage purposely—gardeners call such wind-borne, bird-planted, or otherwise unintentional additions "volunteers"—but I know the shrub didn't just *arrive* in our garden without some assistance. When I dig down, I can see my own roots too.

Mom and Dad designed my childhood home—as much as new buyers could design a home in a 1970s Southern California planned community. They chose the cabinetry, the light fixtures, the colors of the tiles and the carpets and all the interior paint. In that house, my parents enjoyed a sense of ownership and mastery neither had previously been afforded.

The bluffs behind that house ran directly into our backyard. A 2020 real estate listing named my childhood street "the most desirable block in Turtle Rock." But in those early days, earth-scarring construction crews' bulldozers and boots left behind cracked desert dirt. The dry ground was as tan as the earth-toned paint the neighborhood council required owners use on all exterior walls. Not much grew there but invasive mullein and the tough Russian thistle that's commonly called tumbleweed—a nineteenth-century European import now iconic of the West.

I was only five when my parents bought that house in Irvine. My own memory was unreliable, so after I finished planting the Garden in a Box starts that June evening, I called my parents to verify a story I'd long held dear. While I sat on our living room floor stretching my sore muscles, I put my parents on speakerphone. I wanted Ray to hear the story too.

"There was only dry clay soil with a few weeds here and there," Dad reminisced. "We didn't do anything in the first six months because we were trying to figure out what to do."

The detachment in his next sentence struck me: "You were supposed to have the yard landscaped six to eight weeks after you moved in." Not "*we* were supposed to have the yard landscaped," but "*you*." Spoken with coldness. As if *we*, our family, had been pushed out of a collective.

I remember reading *A Wrinkle in Time* when I lived in that house. In Madeleine L'Engle's book, an evil force called IT strove to make "everybody exactly alike." Reading the scene on Camazotz where all the look-alike children in all the identical houses bounce their basketballs all at the same time felt like reading a description of my own neighborhood. To enforce its homogeneity, the neighborhood in Irvine had a community council, the equivalent of an HOA. One day in our first year in that house, a community council member knocked on our door. When Mom and Dad answered, the man from the council told them they were in violation of neighborhood policy. They needed to install their landscaping with all deliberate speed.

This wasn't the language the council member used. He didn't say my parents needed to install landscaping "with all deliberate speed." That is language from a 1955 opinion related to the Supreme Court case known as *Brown vs. Board of Education II*. I crowd my mind like I sometimes crowd my garden. Surprising things grow together in a tangle. When Dad repeated the story of the yard in Irvine, my mind filled in recollected language from a decision directing much of what my parents—and I—have been able to do in our lives.

The part I loved best about the Irvine yard story was my father's answer to the white man who represented the community council. "We are very concerned about maintaining the ecosystem," Dad said he told the man. "We are going to leave this as it is, so as not to disturb nature. Aren't *you* concerned about the environment?"

Mom piped into our conversation to describe the representative's reaction, "He was apoplectic!"

"What do you mean by apoplectic?" asked Ray.

"Turning red and all," Mom said.

"He expected us to be intimidated by him," Dad continued. "He wasn't expecting us to respond like that."

"The house was brand new, for heaven's sake," said Mom. "We weren't the only ones."

Dad said, "We were pretty much the only ones."

When he stood in his new doorway defending his choices, my father would have been in his early forties—five years younger than I was when I asked him to recount the story. But I look at photos of him from that time and he already looks much older than I do. His first forty years must have been more exhausting in many ways than mine.

Later in the conversation, I asked if Dad thought race had anything to do with that encounter. "I think race always matters," he said. "We were one of only two African American families in the neighborhood. And I'm sure it played in his mind. 'We let them in and there goes the neighborhood.'"

Taking her turn in the councilman's voice, Mom said, "'You can't do that! Natural ecology!? No! We have rules!'"

"They sent a letter stating we were past due, and we needed to present plans immediately," said Dad. But my parents didn't work any faster after the council member's visit than they had worked before. "We went almost a year before we put our landscaping in."

I love that story for my father's creative defiance. As if, on the spot, he resorted to the line about caring about the ecosystem as a quick-witted shield against intolerance.

But that assessment of the story reduces my father's commitment to environmental stewardship. In college, he majored in zoology and minored in botany. The stance he took in the door that day was part of his essential being, not something he dreamed up to annoy the community council. My father consciously exercised his choice to proceed with an attuned awareness of the unique landscape where he'd built his home. To do this successfully meant doing things differently and taking his time. Listening to him tell the story that June evening, I had to acknowledge the quiet patience of his ecologically focused, pride-centered response to the council's demands. That willful defiance built a bountiful shelter in which I grew and thrived.

During the year they left the Irvine yard as clay and tumbleweed and mullein, my parents stayed busy. Dad drove to a neighboring city and the hospital where he served as a physician. In the evenings and on weekends, they planned the yard. "We found plants that were more drought resistant, didn't take a lot of watering. We had people out, landscape consultants. We learned the light play at different times of year. Found plants that were easier to grow in that semiarid climate. We made appropriate use of the time," Dad said. "But he didn't need to know that."

Our neighbors in Irvine mostly landscaped with sod patches, daylilies, and boxwoods, making little northeastern estates even in the farthest reaches of the West. Only the occasional hibiscus, agapanthus, or fuchsia, originally imported from the Caribbean and the Mediterranean, nodded to Southern California's temperate ocean-breeze weather. Like the people who designed it, our yard looked different from the others around.

Once, in 1980, Dad knelt in the front of our house, working on the drip irrigation system. Though the idea of drip irrigation to reduce water waste in a home garden wouldn't gain popularity for decades, my parents worked, all those years ago, to make a landscape conducive to survival in a difficult climate. As Dad dug, a white woman came up the walk. He looked up. When she didn't meet his eyes or say anything to greet him, my father kept digging.

The woman rang the bell. Mom opened the door and the woman said, "May I speak to the lady of the house?"

"I told her I owned the house, and she turned beet red," Mom said. "I laughed until I cried."

Ray laughed too. A loud, true laugh that came from his belly. He swiped his forefinger just below his eyes. He asked why, after a decade in our family, he had never heard this story.

"Part of what happens," said my dad, "is you just get tired. When you have one hundred and fifty years of these insults—your mother and I, we have one hundred and fifty years of these between us at this point—if you spent all your time thinking about them, you'd be immobile. Because there are so many of them, both gross and subtle. If you talked about them all the time, you'd never think about anything else."

Despite insults that so frequently suggested they didn't belong, my parents found ways to keep living, keep prospering, and keep laughing. They filled most of their yard with pollinating plants more useful to the ecosystem than dirt or pebbles or turf. But they also installed a swirled swatch of lawn where my older sister and I could cartwheel, jump, and run free. The garden they created, the one I grew up and thrived in, had a wisteria arbor I loved to rest beneath. Birds nested there too. In the backyard, encircled by a wide raised ring of landscaping brick, a willowy-branched pepper tree produced little red pods we could cure and use in the kitchen. I liked to hop onto that short wall and run around and around the ring, perfecting my balance.

We had two fruit trees, each of which should have produced three types of fruit: peach, plum, and nectarine. Though the peach grafts never bonded, we loved season after season of backyard plums and nectarines. My parents planted marigolds and California poppies. Tomato vines and California honeysuckle. Always, something bloomed. Even a jade plant near the front door, gifted to my parents soon after they married. For more than a decade, they carried that jade in a pot from state to state, university to university. When they put it in the soil in California, it thrived, flowering regularly.

I hope for something like that perpetual blossoming in our Colorado garden.

My friends Mary Ellen and Joseph planted a spectacular garden here in Fort Collins, but the house whose yard they beautified did not belong to them. After their landlord sold the property and informed them that they had to move, Mary Ellen and I dug up perennial roots and rhizomes to transplant into my yard. I found a few strands of her graying blond hair wound around some allium bulbs when I buried them. I put her hair in the ground, too, integrating my friend's DNA into our soil—nutrition for the plant and a reminder of a woman who loved them.

Mary Ellen and Joseph left their old house a week after the late-summer 2019 day when Andy cleared the sod from our south lawn. Months before the windy October morning when the big white truck delivered the piles of soil and mulch we'd spread over the bare patch. The transplants I brought from their garden would dry out and die if I left them out of the ground too long, so I planted them in the south plot before the prairie project was fully prepared.

The day we spread the new soil and mulch, I protected Mary Ellen's plants from the chaos by covering them with empty ceramic pots left over from the begonias, petunias, celosias, and other annuals I keep on the patio during warmer months. I intended to uncover the protected transplants after we put the planting materials down.

Ray hired four college students to help me spread the soil and mulch on the prairie project that October Saturday. He knew this would require even more manpower than the two of us put into securing the piles of planting material in the street and driveway earlier that week, but once the workers arrived, he figured he'd done his part. He left on his road bicycle. He desperately wanted to get in a long ride before winter.

Part of living successfully in community means accepting that people are always motivated by conflicting needs and desires—much as the blue spruce in the corner of our yard needs many things: root space, and water, and birds who thread between branches to open room where light and air can enter the thicket near the trunk. The spruce needs these things, not in some orderly succession, but all at the same time. For the most part, I have no control over whether the spruce's needs are fulfilled. But I can check the irrigation system to make sure it's not delivering too much or too little water, and I can be careful not to crowd the tree.

To glimpse the prairie project from the sidewalk, passersby have to peer around the spruce's lower branches. When we pulled the sod, we left a fifteen-by-fifteen-foot stretch of grass around the spruce. I chose not to extend the prairie project's plantings all the way to our south sidewalk. I didn't want to harm the tree's roots with the sod cutter. I didn't want new plants too close to the tree, competing for water and nutrients. I love that tree, and the tree, I knew, needed plenty of space.

Six hours into that October Saturday, when Ray returned from his bike ride, the hired workers, Callie, our friend Megan, and I were still preparing the south plot. We had already placed layers of weed-suppressing cardboard and all the soil. A section at a time, we spread newspaper meant to serve as an additional weed barrier. We wet that newspaper and then covered it with a carefully calibrated layer of mulch. We moved methodically, not quickly.

Sweaty from his ride and worried about the onset of the night's predicted snowstorm, Ray told the young men how to speed up the process, contradicting the guidance I gave all day. As a result, they spread the mulch too thickly. I managed to find and remove a few of the ceramic pots that evening, but not all of them.

In early April 2020, before the threat of spring snow passed, Mary Ellen's bearded iris shot sharp green triangles through the thick layer of bark chips. Other transplants followed. A few patches of columbine greens, in tri-leafed bunches like clover. A stand of grasslike leaves that eventually revealed themselves as the allium, a cousin of onion and garlic. Two more green clumps turned out to be bellflowers. Their purple cups opened to the sun the last week of May. But months of snow had compacted the mulch, and now I could see six pots the rushed workers had buried deep in the planting material. I dug and tugged and wrenched the ceramic hoods away. Mostly, nothing but dirt filled the exposed cavities, though April warmth did pull two sprouts toward the surface. Wispy ghost-white stalks stretched toward the hope of sun. I tamped soil around the chlorophyll-deficient stems and hoped for the best. Neither ever produced any greenery.

As I cleared thick patches of mulch for ground-nesting bees on the June 2020 day that I planted the Garden in a Box starts, my shovel struck another two buried pots.

Shoots of rage flared up in my heart. Ray's haste undermined my efforts to build a carefully planned garden. His hurried strategy smothered my friends' gifts. In my head, I repeated the words I'd flung at my husband that October 2019 evening: *"You can't disappear for hours then come home and try to control what happens in my garden!"*

I had to stay in the yard a long time—collecting myself. I pulled excess mulch from our yard and spread it where Bob wanted it around their roses. I shoveled and sweated until I no longer felt the need to scream. Then, I went inside to call my parents.

I don't want our garden to harbor rage.

A few days after I finished planting the prairie project, our friend Tim came to the backyard for a socially distanced visit. His daughters wore masks that highlighted their eyes and the long straight blond hair framing their faces. Callie and his girls took turns running between the front and back yards, hiding a ball for a scavenger hunt. Months of separation had been hard on the children, but I watched my own daughter's resilience grow, her imagination alongside it. She learned new ways to keep her mind busy. New ways to make herself smile.

Ray had returned from another long ride. He still wore his

cycling gear, and I could see the salt on his face from hours of sweat. "I track him, you know," I told Tim, showing our friend the app on my phone that records Ray's route, pace, and heart rate while he's out on his road bike.

"Of course you do," said Tim.

That spring, white men murdered Ahmaud Arbery, a Black man who jogged on what his assailants considered their road. Around the same time, a white woman cut short Christian Cooper's birding outing in Central Park when she called 911, accusing Cooper of threatening her safety. Demands for reform and retribution charged the summer air. Tim expressed excitement about the building momentum of resistance. "Have you seen how many antiracism guides hit the *New York Times* bestseller list this month?!"

I'd seen that. I'd seen all of this before. In 1991, when video footage caught Los Angeles police officers brutally beating Rodney King. The clip played over and over during my last months of high school and first year of college. In 2012, when a neighborhood vigilante shot Trayvon Martin, around the time I weaned my own child. After Michael Brown and Eric Garner in 2014, when Callie was still in preschool. Garner's youngest daughter was three months old. After Philando Castile in 2016, when a four-year-old sat in the back seat of a car while a police officer fired round after round after round after round after round into her father. I told Tim that I doubted much would change because of some media coverage, some marches.

One of the intentions of intolerance is to redirect my energy, to require me again and again to prove my worth and defend my right to exist. Understanding, as my father said, that if I focused every conversation on racialized violence I could spend my whole life immobilized by fear and anger, I said to Tim, "I've been focused on my garden."

Exhausted—and incredulous that my voice could do much to change the violently racist trajectory of our nation—I chose to focus on what grew from the soil in my immediate surroundings.

Tim wouldn't stop. "This time looks different." More white people involved themselves in the protests, he insisted. "It's true," he conceded, "your voice can only go so far." White people had to buy into necessary systemic changes. "It's also true," he said, that peaceful revolutions are slow. But that doesn't mean they aren't effective.

Shifting the conversation back to Ray's rides, I told Tim, "Ray's tracker tells me if he stops moving." The GPS device is sensitive. "He has learned to text me and let me know what's happening." Imitating Ray, I said, " 'I'm stopping for a snack. Everything's all right! Don't worry.' "

Ray stands nearly six feet tall. He's broad, with skin dark as rich loam. Salt-and-pepper dreadlocks fall to the middle of his back. Even in full road cycling gear, with gloves and a helmet and cycling glasses, Ray is hard to miss. Tim and I pictured him pulling to the side of the trail to peel a clementine and drink some sports gel. First taking a moment to help calm my fears, directing his large, gentle fingers to text me.

Tim laughed. "You can never be too safe."

A recent national study reveals that Black cyclists are 4.5 times more likely to die in an accident than white cyclists. Law enforcement agents in cities around the country cite Black cyclists at a grossly disproportionate rate. Though the population in Tampa, Florida, is only 25 percent Black, Black cyclists accounted for 80 percent of the city's bicycle citations in 2015. In a nine-month span in 2017, Chicago law enforcement agents issued 321 citations to cyclists in Austin, a predominately Black neighborhood, and only 5 in predominately white Lincoln Park. Disproportionately violent interactions with police, for minor incidents, or for no cause at all, had people across the nation protesting the deaths of Black people like George Floyd, Breonna Taylor, and—an hour away in Aurora, Colorado—Elijah McClain.

"Nope. I'm not going to let him ride for hours without my knowing where he is," I said. "He rides through LaPorte sometimes, for heaven's sake!"

The *Denver Post* ran an article in 2012 identifying seven Colorado-based hate groups. According to the Southern Poverty Law Center, this number tripled in the subsequent eight years. One group that's consistently on the list maintains headquarters in LaPorte, a town adjacent to Fort Collins. It's beautiful out there, with red rocks and limestone escarpments, tree-lined river runs, and access to miles and miles of gorgeous, groomed trails. But to get to those trails, Ray has to bike right in front of the group's main church building.

"I try not to go there if I can help it," said Ray.

We could hear the game the girls played in the yard. Three of them hid a ball. The fourth searched for it. The quarry seemed to be underneath the giant rhubarb leaves, near where one of our yard rabbits, Bun, hollowed out a space for his body.

"You're hot," the young guides screamed. "Getting hotter!"

The girls' chorus punctuated our conversation. "No! No!" they warned the searcher. "Now you're getting cold. Turn around! You're getting really cold."

Given spring 2020's stay-at-home orders, the neighbors we saw most frequently were mountain cottontail rabbits. From my standing desk at dawn, the only unfettered time I have to write, I watched some of these rabbits through my study's window. Dee and J., the women across the street, have a sloping stretch of lawn that serves as a buffet for a few of these rabbits. There's probably fourteen hundred square feet of turf over there. Cottontails are crepuscular animals, feeding most leisurely at dawn and dusk, and on these mornings, the rabbits break their fast with near abandon. Well-kept juniper bushes line the walkway to Dee's and J.'s door. When a person walks a dog on the sidewalk, I watch whatever bunny is munching hunch into the grass to look more like a rock than a living being. Then the rabbit zigzags across the lawn in the diagonal pattern evolutionarily designed to fluster predators, and dives into the juniper bushes.

Mountain cottontails are solitary rabbits. I rarely see one closer than twenty feet to another, though their numbers and density increase with more favorable resources. In 2020, our neighborhood boasted a fluffle of bunnies. In the front, back, and side yards, we saw tightly alert bodies munching on grass. Their fur fluffed when they caught scent of us. Then came the racing away. This species doesn't dig deep holes, so to call the ones in our yard a warren, another collective name for rabbits, would be imprecise. Rather, mountain cottontails nest in existing shelters: bushes, woodpiles, tight spaces under decks. A bunny who ate Dee's and J.'s grass lived in the juniper bush below their dining room window. Pebble, the small rabbit who lived in our backyard, favored the woodpile in our southeast corner.

First there was Bunny, who was probably several different bunnies we confused as just one. We mostly noticed Bunny crossing our front yard from the northwest juniper bush to the spruce tree that dominates the southwest corner. Mountain cottontails have brown fur full of white highlights, and their bellies look to be closer to a solid white. Their ears are not foppish like the ears of a jackrabbit. Rather, they seem proportional to their bodies, suited to the work of listening for danger while still blending into their surroundings. They enjoy the dandelions as much as the grass, and otherwise leave most of what I grow alone. A bunny or two has always lived in our yard, but because 2020 kept us in the house, observing, the bunnies revealed their individual personalities as we took the time to pay attention. Out of an aggregate, Bunny, came Lily. Then Bun-Bun. Then Bun. Then Pebble.

Callie named them all. She insisted we call the first Lily. I wanted to call her Dewlap, to commemorate the feature, a fat roll of fur below the chin, that differentiated her from the rest of the fluffle. But Callie didn't think it right to identify somebody solely by the physical trait that set them apart.

I thought Lily was a *he* at first. The rabbit looked like one of those fat, rich, lace-collared men in sixteenth-century European paintings. The fur roll seemed to make it difficult for the bunny to lower his mouth to the grass. I looked up what was wrong, and what I learned changed a great deal about a lot of what I see.

In ways I can comprehend if I look more closely, many of the beings in our yards are gendered. Take the blue spruce and the juniper bushes, for instance. Spruces grow male cones all over the tree, but also produce female cones two stories up, in the top 10 percent of blue-tinged branches. I hardly noticed those big female cones, which have the same size ratio to the male cones as a human egg to a human sperm. Or I hadn't understood that the different sizes represented different gender expressions on the same tree. The spruce is not the *it* I often used for a label. The tree wasn't entirely a *he* or entirely a *she* either.

There are plenty of examples in the living world of beings who are neither entirely *he* nor entirely *she*, but we don't have much capacity to name this in English. The Hawaiian tree snail Lonely

George died on New Year's Day 2019. His name echoes that of Lonesome George, the Pinta Island tortoise of the Galápagos Islands who died seven years prior. Like Lonesome George, Lonely George had no one with whom to mate. These tree snails can live longer than twenty years in the wild. Part of a captive-breeding program, Lonely George survived for fourteen years, outliving the rest of his kind by nearly the entire measure of his solitary life. Unlike Lonesome George, though, Lonely George wasn't really the last *male* of his species. The name Lonely George reinforces a gender binary—imposing a masculine pronoun on a species for whom such constructs are misleading, diminishing, and untrue.

I dishonor the *Achatinella apexfulva*, even as I work to honor the memory of the species, by giving the last such an incomplete name. *Apexfulva*, the latter of Lonely George's binomial names, speaks to the yellow tip of the snail's shell, but reducing a life to nouns that describe appearance and utility erases centuries of context and connection. Lonely George: the last representative of what was the first species of Hawaiian tree snail described by Western science, via shells in a lei presented to Captain George Dixon in 1787. In the days before the arrival of Europeans, *aliʻi* wore leis strung of shells left behind by snails who'd completed their life cycles. Living in trees in the mountains, the snails dwell in a realm connecting Earth to the heavens. To chant to the living snails, or to wear carefully gathered shells, puts a knowing person in a reciprocal relationship with the Earth and the ancestors. This is a lifeway that's vastly different from one that thought the snail's shape and coloration pattern made it attractive for souvenirs. The name *George* speaks of the first white man to hold such a creature. Or, truly, to hold shells removed from living beings and living histories. To name this snail *Lonely George* reduces a whole part of the earth to a solitary *endling*,

driven to isolation and extinction by the introductions of destroy-
ers of communion and interconnection. "That's the difference between a lei and a trinket," said Dr. Lehua Yim when I asked for help
understanding the history connected to this land snail's disappear-
ance. "You have to ignore Native Hawaiian lifeways to get to that
trinket."

The name *Lonely George* speaks of another shelled being, one
who was already lost before many of us found him. Cascades of catastrophe driven by an absence of regard for ecosystems on which
these beings—all of us—depend lead to solitude and peril. Their
status as the last hope for their species made these lone survivors
recognizable enough to some of us, important enough, to grant
them English names. *Lonely George. Lonesome George.* Before their
near eradication, and then eradication, we labeled *Achatinella apexfulva* and *Chelonoidis abingdonii* with binomial nomenclature and
functional descriptions—Hawaiian native land snail or Pinta Island
tortoise—to mark them as separate from humans. Only when the
world had almost lost the last did caretakers bestow on them the
familiar name *George*. But when we spoke about refusing the harm-
ful colonial obsession with extinction narratives, Yim said, "Maybe
this pūpū's direct line dies out, but we go horizontal in our relations
as much as we go vertical. It's a choice to treat this as a final end."

Kāhuli ("land snails"), *pūpū kuahiwi* ("mountain snails"), and the
more Hawaiian poetic terms *kani ka nahele* ("voice of the forest") and
pūpū kani oe (which Yim says is a phrase that "gives a tiny glimpse
into a different social world where the kāhuli continue to be loved
and regularly sung to"): these are some of the Hawaiian names for
members of the family Achatinellidae. "The Native Hawaiian in-
timacy with these snails can be heard in the linkage of their name

with the sounds of human chanting—and also the sound of a baby's wail," said Yim. Though the forests of Oahu once resounded with the songs of tens of thousands of snails, the Hawaiian people and language are still here, some forest snails are still here—cared for and protected by people who refuse settler colonial erasure. Also still here are the imperatives to listen and care. If I seek to connect myself with the profound possibility of true communion on this planet, I must be careful to not shrink the story of this life—of so many lives—to labels as imprecise and incomplete as *lonely* or *insert death-doomed name here* or an uncapacious pronoun like *he*.

Arborists call trees like the blue spruce, who carry both male and female cones on one plant, monoecious. The Linnaean term combines *mono* and *oikos* and means "single house." If granted the space of a forest, such a tree could self-perpetuate. On the other hand, the junipers in our yard are dioecious. They require differently gendered plants for pollination. The junipers in our yard aren't ungendered as the word *it* suggests. *It* is yet another limiting word that erases history and context and the realities of how living beings operate in community. In fact, all but one of our bushes are male. The reason for this tells a lot about what my culture values and does not.

Starting in the 1950s, the USDA recommended planting male trees and shrubs in municipal areas. Nurseries grafted and cloned more male plants for residential buyers as well. Male trees and shrubs don't shed messy fruit, seeds, and seedpods that litter sidewalks, yards, and streets. They do, however, produce a great deal of pollen.

One day in late April, when Callie and I took a break from her remote schoolwork to play wall ball against the garage, the ball landed in the juniper by the driveway. A great puff of what looked like smoke flew up from the bush, mesmerizing Callie. Standing closer, we shook the juniper again and watched the bush release a cloud of pale-yellow pollen. Some researchers believe that the worldwide rate of environmental allergies has increased by a power of magnitude over the last seventy years. This is partially because of the way humans have engineered the sex of the trees and shrubs around us. Now that there are significantly fewer female trees to absorb pollen into what would become messy fruit, and now that the natural balance of male dioecious plants to females has tipped overwhelmingly toward pollen-producing male plants, many people suffer from an increase in seasonal allergies.

Betraying my own flawed assumptions and biases, I misgendered many of our yard bunnies, including the bunny with the thick neck fur, the dewlap. She was female. If not already pregnant or nursing, the roll of fur below her chin prepared her for such a condition. From this extra fur, a mother rabbit plucks the soft lining of her nest.

I haven't seen the inside of Lily's nest. Some things should stay private. Some things, like the birds' nests tucked deep in the blue spruce's high boughs, are designed to keep out those who might unwittingly, or purposefully, do harm.

Bun-Bun lived under the juniper and wintercreeper bushes in the dooryard, near that milkweed that had concerned my mother. When Bun-Bun was tiny, he stayed near the bushes and ate turf and the occasional dandelion. If we came outside, he'd dive under the evergreens' cover. Callie left offerings of torn clover and ripped grass near his nest site, but Bun-Bun refused them. When Callie and I first noticed him eating grass by those bushes, he was tiny, the size of a fist. We thought he might have lost his mother.

Research assured us that Bun-Bun was not an abandoned baby. PETA's website says, "A good rule of thumb is, if you have to chase a baby rabbit to catch him or her, the rabbit is fine."

Mountain cottontails, I learned, are altricial, born blind and without hair. So as not to draw the attention of predators to their helpless young, mother rabbits leave the nest unattended most of the time. The kits feed from their mothers until they are ready to go out on their own. And then they go out on their own. A mother rabbit can have between two and five kits per litter, and between two and four litters a season—multiplying "like rabbits." But, in 2020, we saw only one juvenile rabbit in the yard at a time. First Bun-Bun, then Bun. Then the one Callie called Pebble because she was so small.

We first spotted Pebble near the river rocks clustered along the fence in the backyard. She preferred the woodpile in the back corner. Initially, she hung around the shallow part of the lawn underneath our remaining aspen trees. We use that section of the yard the least and, as we keep no pets, Pebble was generally unmolested there.

She grew braver as she grew larger, often venturing deeper into the back lawn, near Callie's geodesic climbing structure, or close to the rhubarb patch that we can see while washing dishes. Bun had often wallowed there, napping in the shade of the rhubarb's broad leaves.

The average lifespan of a mountain cottontail in the wild is under two years. They get picked off by predators, privation, traffic, disease. Though they belong here, this neighborhood is inhospitable in many ways. One night before dinner, we saw a house cat stalking near the bushes where Bun-Bun sheltered. Free-ranging cats kill more than two billion birds a year in the United States, and as many as twelve billion mammals: mice, squirrels, chipmunks, and rabbits. Ray chased the cat away. But after that evening, we never saw Bun-Bun again.

We didn't see Bun or Lily either.

Soon, we spent our energy rooting for Pebble.

Someone, some bunny, had chewed the newly planted prairie mallow to the nub by August. They gave the bluestem grass quite a trim. Mountain cottontail rabbits—Bun-Bun and Bun and Lily and Pebble—enjoy prairie mallow, wine-cups, and little bluestem grass. All plants native to this landscape, just like these bunnies. The allium must not taste good, because sometime each early summer I find a straight cut allium—stem and flower—lying on the ground like part of a discarded bouquet. Sometimes I bring the stem and its spike-rayed, scentless blossoms inside to beautify my table. Sometimes I leave them right where they fell, until their seeds darken into a mature blackness I can spread for more flowers the next year.

Though the bunnies seem to always be eating, of all the plants I put into the yard, the prairie mallow, the wine-cups, and the little bluestem are the only ones they nibble to the ground. Someone suggested I set up fencing to keep the rabbits out of the prairie project, but the divisive impulse in that solution made me recoil. My answer: plant more mallow, bigger bluestem.

Among all the things we grow here, let the rabbits make themselves at home.

Juvenile mountain cottontail

By the time I got close to the body, the eye sockets were empty of everything but small black bugs. I didn't know who it was. Lily, or Bun-Bun, or some other. A patchy pelt fell off the carcass in tufts. This was the first week of August 2020. It didn't wholly surprise me to find death so close to our door.

I saw the dead rabbit while walking toward the rock garden that grew under the shade of our northern neighbors' Norway maple. Years before, I read that Ancestral Puebloan people around Mesa Verde sited farms in places where they saw sagebrush and rabbitbrush thrive. Those plants indicated accessible water. When I started a new garden in that rock-filled eighty-square-foot expanse, I looked for clumps of bindweed, prickly lettuce, cloverish black medic, and herb-scented creeping Charlie. In hour-long spurts during a handful of fall 2015 and spring 2016 afternoons, I cleared patches of the undesired vegetation, plasticized fabric, and river rock that spread between the fruiting juniper and a bed of hens and chicks (*Sempervivum tectorum*). Where I removed the rock and opportunistic plants, I uncovered drip lines installed by the house's previous owners and sections of well-hydrated soil the lawn's existing sprinklers reached. I left the hens and chicks. Regardless of access to water, those low-growing succulents appear to have thrived in that northwest corner of the yard since the Oligocene.

I laid down fresh soil, seed, and starts, and by 2018 we had white, pink, and fuchsia sweet William blooms that looked like tiny lace doilies and smelled like an elegant great-aunt's perfume; Sonoran Sunset hyssop with dusty-mauve trumpet flowers; bright orange California poppies; purple Nuttall's larkspur; yellow-flowered rabbitbrush; pink and purple bachelor's buttons and the golden crowns their deadheads leave behind; swamp milkweed; multicolored snapdragons; and a crop of bird-scattered sunflowers much shorter and later blooming than those towering in the dominant patch near the driveway. As I hoped, all these plants grew well on their own, but on particularly hot and dry days I sprinkled them with rain barrel water in the evenings.

Watering can in hand, my attention focused on the nodding yellow sunflowers and the purple hyssop blossoms bobbing in the evening breeze, I walked barefoot that August night.

I nearly stepped on the body.

Inside, dinner simmered. The rice cooker had only eight minutes left. I didn't have time to bury a rabbit.

Three days before, our local school board had decided to "conduct all instruction remotely for at least the first quarter of the academic year." As the superintendent's automated message played through my phone, I pictured Callie beginning a frustrating fifth-grade year at our kitchen table and spiraled into cavernous despair.

The morning after I nearly walked on the rabbit, I was scheduled to drive up the Poudre Canyon to meet the family of my mother's chosen sister, my aunt Mary. Finally, we had set a time to scatter her ashes in a highland meadow she loved.

My introduction to the best things about Fort Collins came through Mary, who had lived here since the 1970s. But four years after we moved to town—nearly to the day—her big heart gave out.

One of her children kept Mary's urn in the house for three years. A container for grief. It took all of us who loved her a long time to get over the initial awfulness of not having her near.

August heat pressed heavily around me. It threatened to rain most afternoons that summer but rarely did.

As the high-summer weeks of late July and early August settled over us, Pebble grew so large Callie wanted to call her Dwayne "the Rock" Johnson. We hadn't seen the bunny since the superintendent's August 5 call, but that July Pebble had spent time in the middle of the backyard where we could easily watch her nibble the grass. Or she secreted herself near the raised kitchen beds we put up along the north fence. Once, when I went out to water the vegetables, Pebble ran from behind a pot of zucchini vines, jostled from her sanctuary in the squash's broad green leaves.

For several years, I tried to grow vegetables among our flowers. Around mid-July, as the summer garden really perked up, pumpkin vines would grow where we tossed discarded jack-o'-lanterns and pumpkin pulp for tree squirrels to nibble the previous fall. But those volunteer vines didn't fruit until September. The young squash wouldn't have a chance to harden off before the first killing frost. A pumpkin can take 120 days to begin to mature, and our growing season is only three weeks longer than that. Savvy gardeners start vegetables inside months earlier than our winter soil allows for spontaneous sprouting. Starting in March instead of May extends the viability of plants. Some gardeners install hoop houses and other heat-trapping devices on outdoor plots in early spring and again in the fall, but I didn't pay that much attention to our vegetables.

Sometimes, rather than toss the scraps into the compost, I plugged the root ends of spring onions into the dirt along the east fence. Those yielded enough new onions to enhance a salad. But even after the three perennial strawberry bushes established themselves in the irrigated backyard beds where I sited them, they bore berries so small and sour they hardly seemed worth the harvest. We left their fruit for the birds.

Even the new raised beds produced spotty yields in 2020. I put the peas in too late and they wilted when the heat came on in late June. I harvested eight small tomatoes, two zucchinis, a cucumber. I got enough lettuce for three weeks of salads before that and the cilantro bolted, the stressed plants sending up flowers as the once sweet leaves turned bitter in the hot August sun. The chili peppers thrived in the heat, but only Ray and I like spicy food. When I cooked with the chilies, I made a separate dish for Callie. White cabbage butterflies—but only those butterflies and their voracious, slender green larvae—enjoyed plenty of cauliflower and kale.

A lot of things died that summer.

I kept searching for what would survive.

I'd begun a more serious study of Indigenous foodways, trying to align our diet with the landscape of our life. But I hadn't gathered the purportedly tasty and nutritious tips from the blue spruce during the spring harvest season. The previous fall, I had collected juniper berries from unpruned fruiting branches. I ground the berries with a mortar and pestle and used the peppery paste to season some pronghorn steaks gifted by a hunter friend.

"You don't have to repeat that," Ray said, expressing uncharacteristic displeasure at the meal I offered.

I enjoyed the woody, peppery flavor the juniper added to the sage-strong wild meat, though it took me an hour to harvest and prepare enough dark purple berries for the rub. Juniper berries are highly compacted cones that take two years to mature. Under my pestle, they grew grainy. To create a relatively smooth paste, I worked them in the mortar until my hand grew sore. But from the look on Ray's face as his tongue removed sharp bits of juniper berries from between his molars, I realized I had stopped grinding the berries too soon. Cooking with foraged juniper required a lot of effort for a small reward. I returned to the mostly Mediterranean spices in my cabinet the next time I cooked pronghorn, blending the wild flavor of the local meat with familiar imported seasonings. A bridge between what is new to me and what I've known.

From January through April 2019, Ray and Callie and I lived in Tennessee, where I served a semester as a visiting professor at Vanderbilt. The weather in Nashville felt different from Fort Collins. We didn't know how to describe it. "It's cold," we told friends. The thermometer didn't seem to register the intensity of the cold as we felt it. It could've been the chill-trapping dampness of that region along the Old Natchez Trace that kept us shivering. But I think the acute awareness of all the familiar things that no longer surrounded us made us feel cold. The wind sounded different in the Tennessee house. The trees didn't look the same. It took me all those months in Nashville to put a word to the feeling. But I did: *lonely*. We were lonely. Raw and alert. We never settled in.

After our four-month stay, we spent several days driving back to Colorado. We stopped two days in Memphis. There, we visited Stax Records, where Otis Redding, Wilson Pickett, and Isaac Hayes cut hits. Then we visited the National Civil Rights Museum at the Lorraine Motel. Winding galleries took us past Green Book guides that compiled information about places, including the Lorraine, where Black travelers in the 1930s through 1960s could hope to stop safely for a meal or a night's rest. One room held a bombed-out bus from the Freedom Rides. Displays replicated picket signs held at the sanitation workers' strike that brought Dr. Martin Luther King Jr. to Memphis, and the Lorraine, in April 1968. The last major stop in that building took us to the motel room where Dr. King stood on the balcony in the final conscious moments of his life. The tour then directed us across the street to the boardinghouse bathroom from which Dr. King's assassin, James Earl Ray, fired his murderous shot. As we walked away across the open plaza in front of the Lorraine, spring sun warmed the restless southern air. The trees' unfurled leaves filled our ears with their long, low, unyielding shiver.

We drove next to Saint Louis, Missouri, and visited the Gateway Arch, a 630-foot engineering marvel erected to celebrate America's westward expansion. Before riding to the top of the monument, we toured the lobby's museum. I stood for a long time in front of a display about a four-decade stretch when the city was governed first by France, then by Spain, again by France, and then by the United States. With each shift in power, sometimes practically overnight, the women there enjoyed limited, then far more extensive, then again limited, and then no personal property rights. Each woman remained the same person she'd been the day before. Yet her life and prospects changed completely. Her reality shaped less by what she saw than by how she was seen.

From the top of the arch, the three of us looked toward the courthouse where Dred Scott petitioned for his freedom in a case culminating in the 1857 Supreme Court decision that declared a Black man "had no rights which the white man was bound to respect." That decision remained precedent for generations. It laid the groundwork for the racialized discrimination and violence howling through America's nineteenth, twentieth, and twenty-first centuries. Events that happened 60 or 160 years ago direct our present. If I could leave the past in the past, maybe it wouldn't haunt me. But I stood in the threshold to the Lorraine Motel room where Dr. King talked with his friends, framed in that moment of peace before the peril. I walked in the same courtroom where Dred Scott argued for his rights as a man. I touched that wood. I felt the same ground beneath my soles. No matter how many years have passed, no perennial in life's garden roots more deeply than history.

An animated map display in the Gateway Arch Museum showed the locations of more than five hundred treaties made between the US government and various Native nations since 1776. The United States broke every single one. This map represented 1,510,677,343 acres seized via broken, unratified, corrupt, or annulled treaties. Spanning an entire wall, it looped animation of growing and shrinking claims again and again and again. I watched six times—my own heart growing and shrinking at a predictable pace along with this representation of power, property, appropriation, and privilege.

That night, we ate dinner at a friend's house in Iowa City, Iowa. I've known this friend since we were both fifteen. I regularly sat at her family's table as one more welcomed child in their bountiful Irish Catholic clan. She'd moved into a new house a couple of months before we arrived. While dinner simmered, we walked around her little yard's emerging fairy garden, anticipating the blooms spring might yield. What a relief, after all the haunted histories of the preceding days, to sleep well in that house, filled with love.

Our last leg home took us through Kearney, Nebraska. There, the Great Platte River Road Archway Monument celebrates the path that led Euro-American settlers—some from the same famine-blighted Irish county my friend's ancestors fled—across the American West, following a trail of broken and nullified treaties. A few miles west of our house, a road called Overland Trail memorializes the route. Outside Kearney, we drove toward Overland Trail without interruption.

Somewhere just past the 100th meridian, the landscape opened. With that opening, I better understood the familiar song: *Oh, give me a home, where the buffalo roam . . .* We really saw deer. We saw antelope. And the sky! The less water a basin holds, the larger it looks. Dry air makes the western sky wider and larger and bluer than any we lived under during our months away. Fewer trees in the arid climate mean broader vistas. Without ambient water particles found in humid climates, the sky holds less haze, fewer clouds, so sometimes the sky out west truly *is not cloudy all day.*

We kept the windows of our Honda CR-V closed on that 2019 drive. Still, I could imagine the sweet, snappy smell, more dust than damp, on the breeze. Driving from the Mississippi River basin up to our five-thousand-foot town, we started to feel at home. Callie and I sang together when we hit the familiar landscape of the open range, where the growing season had just started. I saw rabbitbrush and manzanita scattering the plains. Purplish pink blooms of Nuttall's oxytrope appeared like furry snapdragons clustered close to the ground. They reminded me of the little pink flowers Aunt Mary kept in window boxes around the front of her house and on the flower-lined path to her door. Synonyms for welcome.

Our gathering spot to scatter Mary's ashes was a cabin an hour from Fort Collins that Mary had visited throughout her life. There, the weather can be entirely different. I wore long hiking pants, high socks, and my closed-toed Keens. I had a rain jacket in my back-pack, in case a storm suddenly brewed, as it can at that elevation. I brought plenty of water to ward off dehydration.

My mother and father had known Mary longer than any of the rest of us, even Mary's own children. But the pandemic held my parents in Fort Collins because there would be too many people gathered in the meadow. Callie preferred to remember Mary vibrant and alive, not as a can full of ashes. Ray stayed home with her. So I drove alone.

Joining me in the meadow were Mary's two children and their spouses, her grandchildren, a dear friend and her daughter, another close friend, and the cabin's owners. We kept our distance from each other, confined to our individual family units. Four here. Four there. Two. Two. One. And one. Everyone masked. In that high prairie meadow, at eight thousand feet, the sky was that familiar blue expanse of openness. Sage and dust and dry grass and the minty smell of quaking aspen scented the air. The creek didn't sing that day. It hadn't rained.

We took turns scooping ashes from a box-shaped urn, each of us walking to our own spot in the meadow to share words with just ourselves, Mary's remains, and the breeze. The grandchildren tied biodegradable red ribbons on sage and yellow-flowered golden currant. Memorials to Mary waved like sparks through the open field.

Nothing remarkable transpired that morning. There is no reason to tell the story of her scattering except that I cannot—and do not want to—believe Mary is gone. I can't write a book about this place without Mary in it.

One thing Mary fought for while she lived was broad-reaching understanding, and I am grateful there are many lives through which

I have come to understand this place I call home. Driving toward Mary's scattering that August, I kept thinking of a man named Thomas Nuttall.

I learned about Nuttall through the rabbits in our yard, as well as the larkspur in the rock garden that grows in the shade of the neighbors' Norway maple and the oxytrope I saw out the window as we drove home from Tennessee. Nuttall's cottontails—some call them *Sylvilagus nuttallii*. Nuttall's larkspur (*Delphinium nuttallianum*). Nuttall's oxytrope (*Oxytropis multiceps*). The former two named for him, not by him, to honor the wealth of western American species Nuttall brought to the attention of England, Europe, and the relatively new settlers in the northeastern United States. In addition to gathering plants to distribute to collectors, Nuttall wrote, illustrated, and printed several catalogs of North American plants in the early decades of the nineteenth century.

Let me tell you some of the plants named for or by Nuttall that live in my garden. I love the physical and sensorial dynamism these taxonomical and common names point toward: Showy goldenrod (*Solidago speciosa*). *Rudbeckia amplexicaulis*, yellow-petaled flowers with elongated cone-shaped centers similar to but not the same as the *Ratibida columnifera* (commonly known as Mexican hat) that Nuttall also described and which is also in my garden. *Penstemon albidus* and *Penstemon grandiflorus*, flamboyant flowers waving like many flags off tall stalks. *Artemisia longifolia* and *Artemisia ludoviciana*, two kinds of sage. *Mentzelia multiflora* (desert blazing star), and the Rocky Mountain iris (*Iris missouriensis*) I so deeply love.

While journeying through the American West, Nuttall also described the *Yucca glauca* (soapweed yucca), *Ribes aureum* (golden currant), and *Sheperdia argentea* (silver buffaloberry) I see flourishing in some nearby neighbors' yards. Describing Nuttall's influence in *The Plants that Shaped Our Gardens*, David Stuart writes, "Almost every European garden and many American ones still grow many of his flowers."

Though Nuttall never came to what would become Fort Collins, he trekked within 160 miles from here during the summer of 1834. Driving north toward Mary's scattering, from the Fort Collins spur of the Overland Trail and through rocky escarpments and clay soil prairies, I imagined Nuttall traversing similar terrain. Along with the ornithologist and naturalist John Kirk Townsend—for whom science named Townsend's warbler, the white-tailed jackrabbit (*Lepus townsendii*), and Townsend's big-eared bat—Nuttall served as a botanist and naturalist on Nathaniel Wyeth's two-year expedition. They followed the Corps of Discovery's route from Saint Louis, through the Missouri Territory, along the North Platte and Snake Rivers, toward the mouth of the Columbia.

Wyeth wanted a stake in the Rocky Mountain region's beaver trade. With such a claim, he might grow as rich as John Jacob Astor, whose agents bought beaver pelts from trappers at $4 a pound—the equivalent of $120 in 2020—then resold the pelts at a steep profit to hatters. Elegant top hats made of beaver pelts cost six times a London laborer's weekly wages, and monopolies like Astor's kept demand for pelts high. Astor reinvested his earnings into Manhattan real estate, where he further monopolized commerce, and he died the wealthiest man in America in 1848. But three days before Wyeth's expedition arrived at an encampment that settlers eventually named Fort Laramie, a rival expedition set up camp there, lightening their own load of trade goods by selling to local trappers, and ensuring their faster arrival at the next highly desired trappers' rendezvous site. Driving to Mary's memorial, in the direction of Fort Laramie, I imagined Wyeth's expedition trailing the more successful party. Day by day losing hope of gaining fortune.

We spread Mary's ashes near a place called Beaver Meadows, so there must have once been plenty of the amphibious mammals around here. But overhunting for men's hats from the late sixteenth century through the mid-nineteenth century mostly extirpated beaver in this region. Feats of twentieth-century human engineering supplemented and supplanted the changes beaver wrought on landscapes and waterways, on a vastly different scale.

I can't know what Nuttall saw here. There would have been no houses like mine then, nor the gridded farms and ranches such housing developments replaced. The prairie Nuttall walked through probably seemed immeasurably vast. That wide western sky broken only by the fourteen-thousand-foot mountains that must have daunted travelers who trekked out of Fort Laramie in search of the lowest pass.

Though Wyeth's sponsorship provided the occasion and funding for Nuttall to venture on a path few white people had thus far traveled, Nuttall was not particularly interested in beavers. It was the flora of the American interior that fascinated him. Traveling with Wyeth, Nuttall collected specimens and seeds to send to collectors in the eastern part of the continent and England, his original home. The Era of Westward Expansion had dawned, and Nuttall helped encourage its dawning. Englishmen, Scotsmen, and Euro-Americans clambered across the vast continent, naming and collecting and calling everything their own.

Around the first of June, when Wyeth's expedition traveled through the area, four- and six-foot stalks of bushy rabbitbrush are powdery gray. At 105 degrees west and 40 degrees north to 42 degrees north, the longitude and latitude of Fort Collins and Fort Laramie, the clusters of tiny yellow flowers that look from a distance like dried yellow baby's breath bloom closer to late August. Nuttall encountered the blooms somewhere, though, because he named the genus goldenbush. He also named three of its species: *Chrysothamnus depressus* (dwarf rabbitbrush), *Chrysothamnus viscidiflorus* (yellow rabbitbrush), and *Ericameria nauseosa* (formerly, *Chrysothamnus nauseosa*, a.k.a. rubber rabbitbrush).

Even without rabbitbrush blooms, plenty of plants flower here in June. Close to the solstice, snow mostly draws back, and the landscape takes advantage of the high sun for the short growing season. Purple blossoms of wine-cups grow close to the ground. Penstemon stalks and golden rods of solidago shoot up waist high. Shin-high *Argemone polyanthemos* look like a cross between an aggressive thistle and a delicate white poppy. I imagine Nuttall out of his mind with the pleasure of collecting so many botanical wonders and the prospect of naming all he could.

Somewhere on his trek, he must have encountered communities of prairie dogs. The social creatures like to cluster near white-barked, powdery-leafed saltbush (*Atriplex nuttallii*). During our first seven years in Fort Collins, I watched the removal of three prairie dog towns—traps set at the entrances to their dozens and dozens of burrows. Our local paper's naturalist, Kevin J. Cook, suggests that, based on their location in the high prairie and their taxonomic family, the appropriate name for these fifteen-inch-long, buff-colored, chubby, highly social mammals should be steppe-squirrels. But what I call them won't matter if they're gone.

I catch myself giving the side-eye to people who live in new-build communities, as if I'm grandfathered into this ecosystem, and the blood of the steppe-squirrels is on these newcomers' hands alone.

We bought a preexisting house because I didn't and don't want to encroach on the landscape and the lives of the landscape I love. I drive alternate routes to avoid confronting new developments. But many are impossible to avoid.

When I visited Aunt Mary in the 1980s and the 1990s, we took I-25 north from the airport, or detoured off I-80, driving fifty miles south from Cheyenne. On those drives, it seemed nothing existed between our car and the Rockies but open fields. When we moved here, I started seeing tan tanks on these agricultural expanses. I figured they held water until my friend Kate visited in September 2018 while she researched *Quakeland*, her book about seismicity. The tanks, she said, belong to hydraulic fracturing ("fracking") installations. Among the short- and long-term catastrophes they might cause, the sites could lead to increased earthquakes in our area. The view I once carelessly cherished began to terrify me. What I see between the mountains and me: pump jacks, oil wells, drill rigs, flare pipes spouting plumes of fiery gas, and hay walls intended to hide the machinations allowing access to methane buried millennia deep. Like the beaver meadows that garnered wealth for John Jacob Astor, all the beautiful places I love—tall and short grass prairies, high meadows, pocket gardens like my own—have been sites of extraction for more than two centuries.

This part of Colorado: half scenic byway, half mining town. On the road to ski Breckenridge, I pass more resource-extraction sites than I can name. Gravel pits. Gold mines. Selective logging operations. Some relics from another century. Some activated recently. What is it to ski but to extract my pleasure from the mountain? Lead, uranium, the silver that delivered a fortune to Meyer Guggenheim and his sons, including Senator Simon Guggenheim, whose own fortunes bought sizable buildings on three Colorado campuses, including the one where I teach. And their sons in turn, one of whose deaths, at just seventeen, is memorialized by the John Simon Guggenheim Memorial Foundation. That endowment sent me a check that encouraged me to step away from teaching for a year and focus on writing this book. Natural gas, feedlots and slaughterhouses, the seeds and pods and roots that Nuttall carried away to the east and to England, the land on which I've built my home and garden, so much rich bounty here in Colorado. But everything—*everything*—comes with a cost.

When I returned from our memorial for Mary, I traded short sleeves for a long-sleeved shirt and a pair of rubber gloves. I brought the flathead shovel and the oval-tipped shovel from the garage, plus two plastic garbage bags, and I walked toward the dead rabbit's body. Despite the heat, I kept on my mask. The more of myself I could cover, the better. We'd had outbreaks of the bubonic plague, and both rabbits and humans are susceptible. I should have called Colorado Parks and Wildlife to report the death, but that didn't occur to me until later.

Maybe malnutrition killed the rabbit. Maybe a cat. Or a dog. Perhaps some intolerant neighbor set out antifreeze. Another possibility: rabbit hemorrhagic disease virus serotype 2 (RHDV2). This disease, originally from Europe, first showed up in Colorado in early 2020. Transmitted via direct contact with infected rabbits or biting insects, RHDV2 threatens both wild and domestic rabbits as well as other lagomorphs like pikas, marmots, and hares. Onset may be sudden. A rabbit will lose appetite, appear lethargic for a day, then die. Fatalities occur within three days of infection, at a rate of 40 percent to 100 percent. "It's highly contagious," I texted Kate and

Suzanne that evening. "So, yeah, we've gotten to the month in this infernal year when bunnies spontaneously explode."

One way to know a rabbit died of RHDV2 is what scientists call a *bloody show* around the eyes, nose, and anus. But bugs had consumed any such evidence on the rabbit in our yard.

"Nature, red in tooth and claw," I said to myself as I prized a shovel under the body.

Tennyson wrote those lines in the elegy *In Memoriam*. The poem honors Tennyson's friend Arthur Henry Hallam, who suffered a fatal brain hemorrhage in 1833. Around this same time, Nuttall chafed at Harvard, eager to resign his post as curator of the university's botanical gardens and walk out with Wyeth into the western wilds. Tennyson's poem wonders whether hierarchies of intelligence, evolution, and creation's favor can protect any of us from the fickle violence of the world. But I wasn't thinking about all that context, complexity, and coincidence when I said, "Nature, red in tooth and claw" over and over in my head. I thought only of how brutal this world can be. My world. My garden.

Within the week, we saw Pebble again, in the northeast part of the yard. We also began to see an assumed latter littermate of Pebble's. We called this tiny one Chip. The new rabbit hopped in and out of the bramble near the woodpile in the southeast corner. Close to the dooryard juniper, we often saw one whom Callie named Weewi, presumably of the tribe of Lily. A bit older and fatter, one we called Sunny, who used to come in and out of the nest where Weewi eventually sheltered, hung out in the front yard too. The younger bunnies continued to appear throughout the August week I found the dead body. Who the lost bunny was remains a mystery.

Toward the end of that month, I went out to water the raised beds. Pebble flattened into the grass, trying to disappear. "It's okay," I said. "I'm only watering. I'll be past you in a moment." But my assurances did Pebble no good. As I walked with the watering can, I pushed Pebble farther west, until she entered Sunny's range, near the fruiting juniper. The two came within inches of each other, seemingly as alarmed by the presence of another rabbit as by me. Then, both bunnies dashed in opposite directions. I didn't see either again for days.

Territoriality is a human behavior. But not human alone. We know Chip and Wee-wi and Pebble and Sunny by their favorite grazing spots as much as by the patterns of their pelts. When our prairie project matures, with more flowering plants established from early spring to late fall, we plan to put in bee boxes to house native leaf-cutting bees and mason bees. Unlike European honeybees, who fly up to six miles seeking pollen, these local species rarely venture more than three hundred feet from their nests. Solitary, not troubled by a communal hive, they aren't aggressive. For them, territoriality isn't about attacking. It is marked by their disinterest in venturing far from their nests. Such site fidelity puts them in danger. The removal of native plants in a localized area means the destruction of relied-on—often exclusively relied-on—sustenance. Territory-dependent species like these have nowhere else they can survive.

I felt bad about displacing Pebble from her territory, especially after observing her apparent terror at bumping into Sunny in that rabbit's range. I wondered, not for the first time, if my presence contributed to whatever caused the death of the bunny beneath the fruiting juniper.

C. D. Wright's words "what elegy is, not loss, but opposition" resonate so much with my thinking that I used them as an epigraph in my book *Smith Blue*, a collection of poetry focused on environmental devastation. I am always thinking about loss. Opposing it. The rapacious behavior of humans has triggered species loss at a thousand—no—ten thousand times the natural extinction rate. While I talk to the still-living bunnies in my yard, I mourn how my presence puts them in peril.

"Nature, red in tooth and claw." I dug under the lost bunny's body, repeating this line from Tennyson's nineteenth-century elegy again and again and again.

But even with elegy on my mind, I was not prepared for the maggots.

Later that week, I called my friend Sean Hill. We're both parents of only children, and we reach out to each other for companionship for ourselves as well as our offspring. We're both Black writers living in the sometimes-isolating American West. We check on each other frequently. "I don't think," I told Sean, "I've ever seen so many maggots."

"Really?" Sean sounded genuinely surprised. He paused for a long while, perhaps silently recounting all the times he'd seen maggots. "You haven't?"

Tracking Sean's silence, I realized I kept myself from maggots by design. If I maintain a carefully manicured yard, I have a better chance of controlling what, and whom, I look at every day. I will be less likely to accidently confront rot and death and pestilence. The *Cambridge English Dictionary* says, "Rewilding runs directly counter to human attempts to control and cultivate nature." One of the synonyms for *wild* is "out of control." This wild-leaning yard I'm trying to foster is bound to offer many more surprises that might not prove pleasant.

Disturbed by my shovel, the rabbit flipped over. But it did not slide into the plastic bag I opened beside it. The giant cavity exposed on the rabbit's underside writhed with several species of pale grub, snotty-looking larvae, nearly translucent worms. Ants too.

The gruesome mass triggered my gag reflex. I tried to breathe steadily through the mask I wore, but I succeeded only in holding my breath. I scooped the carcass and ten shovels full of the surrounding dirt into the bag, tied off the top with a knot, then placed that bundle inside a second bag. I tied that too. Then I stood still and quietly. I could hear the maggots squirm and munch inside the plastic. A muted, arrhythmic percussion. I tossed it all into the garbage and, though trash pickup wasn't for two days, I dragged the bin out of the garage and down the driveway to the street.

In different circumstances, I would simply bury the carcass. Or I might visit and revisit the exposed body, witnessing the transition from death into more fertile soil. Not long before, I found a dead mouse on the front porch, near the begonia and marigold pots. Ants had already started their work on the body, and after a few days I saw no evidence that a mammal had ever been there. But with the rabbit, I wanted to clear the area to limit the risk of disease spread. I used the double bags as much to contain the bugs around and within the bunny's body—possible disease transmitters should they transform into biting insects—as to hold the dead bunny itself. I covered the dig site with two cubic feet of shredded cedar. When I finished, I soaked the tools and gloves in a bucket of water and hydrogen peroxide, put my clothes directly into the washing machine, and showered. I didn't tell Callie about the rabbit.

In the week after I buried the bunny, a massive wildfire broke out thirty miles away. The pandemic still raged. We didn't want to risk infecting my parents by meeting inside, but the smoke from that fire and others across the region kept us from spending time outside. That night, we decided to eat together in my parents' dining room. Each household sat eight feet from the other. A UV air purifier ran on its highest and loudest setting. Smoke and tension charged the air.

"The world is full of suffering," said my mother. "She might as well know."

I'd followed Mom out of Callie's earshot to ask about a fire. The rising COVID-19 death toll, nationwide protests against murderous police, fires raging across the West—at the dining room table, we'd land on a topic and I'd shunt the conversation in less frightening directions. Mom saw Callie's body tense when something traumatizing came up, but she shook her head incredulously when I changed the subject. Why not be honest with my daughter about the violence that roiled the world?

"I don't remember how it started," Mom continued. I stood in her kitchen, not really helping with the dishes. I didn't want to stand too close. "I just know there was a lot of smoke. Someone tried to light the back porch on fire, and there was a lot of smoke."

Between her sophomore and junior year of high school, my mother moved to Chicago from a small Black community in Kansas City, Kansas, where her dad led a church after the family left Virginia. The church in Chicago that hired Granddad bought a triplex in a Jewish neighborhood on the edge of Garfield Park, installing my mom's family on the middle floor. Older Jewish couples lived above and below them. This was 1958. It must've been the fires all around us that made me ask Mom to tell this story. "It's quite simple, really," she said. "Someone tried to light our house on fire because they didn't want Black people in the neighborhood."

On the ledge by her kitchen window, two African violets bloomed in their pots. One of my parents had placed two breakfast settings for the morning on the small kitchen table, a blue vase of flowers between them. I thought what happened outside the Chicago house was a KKK-style cross burning. But the story Mom told was a different kind of scary. "We never learned who did it," she said. "Some boys from the neighborhood, we guessed. They made it clear they didn't want us there."

Mom turned to the sink, then quickly turned back toward me. "I enrolled in Marshall High School that year. I brought all these top grades from Kansas, but they wouldn't accept them." She'd been in an all-Black school in Kansas, even after the 1954 Supreme Court case *Brown v. Board of Education of Topeka, Kansas* should have ended segregation in American schools. Though she brought straight As on her transcript and continued that success at her new, reluctantly integrating high school in Chicago, administrators told her she couldn't take top honors because she hadn't spent all four years at Marshall.

"They couldn't have a Black valedictorian. Not the salutatorian either. They just couldn't accept that a Black girl could be that smart."

More than sixty years had passed, but I noticed the upright posture of her shoulders—a more alert and hurt poise than when she told me about the fire. As if the disrespect of having her intelligence denied burned more than the fact that anti-integrationists—domestic terrorists—had tried to burn down her family's home. "Just the other day," she said, "I remembered that."

We saw a slew of dragonflies and damselflies that summer. An order science calls Odonata. Narrow-winged damselflies, blue-ringed dancers with long black abdomens banded with brilliant blue rings; orange bluets, whose whole slender bodies seemed spun from bright orange-gold; blue dashers, with their chalky blue thorax and brown-spotted wings; twelve-spotted skimmers; widow skimmers; and the common whitetail skimmer, whose black- and white-spotted wings and thick bodies rested on our yard's echinacea, rudbeckia, or purple coneflowers for minutes at a time. Only once had I seen as many dragonflies as I saw in the yard most summer days in 2020. That other time was in 2015. We sat on my parents' back patio with Aunt Mary—still blissfully unaware that she'd be dead in two years.

Mom and Mary met as undergrads. My mother was the first Black graduate from their central Iowa college. The only Black woman on a campus of seven hundred students. When a classmate's father refused to allow his daughter to live with a Black woman, Mary volunteered to be Mom's roommate. They've been like sisters ever since. For decades, Mom and Aunt Mary each displayed a picture of themselves from their thirtieth college reunion in honored spots in their homes. Flowers and dragonflies adorn the matching frames that circle the two friends. When I see dragonflies, I think of these beautiful women's sixty-year-long love.

Naturalists say most of what we noticed during the pandemic shutdowns had been there all along. We just looked more closely that year—and so we saw more birds, more bunnies, more Odonata, more moments of simple, sweet joy. What a blessing.

It delighted Mary to visit with us in the two years—too short!—we all lived in Fort Collins together. She'd stop talking and look, really look, into our faces. Leaning forward in her chair, she ran her hands down her thighs as if she could hardly contain her happiness. The sound of her palms rubbing the fabric seemed to shout, *Yes! Yes. Yes.*

We laughed together, that evening on my parents' patio, in the easy way of old friends.

Mindlessly, we swatted away the gnats and mosquitoes of summer. Until a moment when we noticed the small pests disappear.

I could practically hear "Ride of the Valkyries," that Wagner tune from the cartoons of my childhood, as a formation of fifteen Odonata swooped over the patio. Nature's assassins on patrol. Dragonflies can eat their own body weight in insects—hundreds of insects—each day. I have no idea where these came from, but they gobbled all the pesky gnats that had not already sensed the Odonata and escaped. We marveled at their ravenous flight.

In Colorado, some mosquitoes carry the debilitating West Nile virus. I'd be lying to say I want them around. But I can't say I want gnats and mosquitoes to entirely disappear from our region either. They serve key roles in the ecosystem. Many of the creatures I love, like Odonata, rely on bugs we call pests.

More than once, as I work to diversify the plants and animals in our yard, I have found myself wondering where my insistence on neighborliness will end. "We haven't seen a snake in our yard," I texted local friends who found one in a planter. "But I imagine that, once the prairie project is established, we'll see them more often." On the gardeners' discussion boards I visit, plenty of planters come to the defense of garter snakes, bull snakes, and racers when people post stories about killing reptiles they find in their yards. Beneficial to a garden, snakes control rodents, bugs, and other unwanted pests, and many gardeners encourage intolerant planters to extend grace.

"The prairie is more biodynamic than any region in the world!" said my friend Taylor on a video call during that pandemic-clouded August. We were catching each other up on our writing and our lives. Taylor asked for updates on the prairie project. He loved the idea of replenishing a small piece of the native landscape in our own backyard. "A hundred acres of prairie can support three thousand species of insects," he said, dipping his chin so a few strands of his ginger hair flopped toward the camera. "Species!"

Through faux tortoiseshell glasses, he made sure our eyes met. Taylor needed me to know he meant types, not individual animals. With three thousand species, the number of individual animals on the prairie would be more than I could fathom.

Taylor and I had talked off and on for months. Really, we'd been having this conversation for more than four years. We talked about the garden Callie would grow up knowing and how it differed from the landscape of Taylor's childhood. We talked about the narratives we hadn't seen published out of the American West. Books we longed for. The Black pastoral. The Black garden book. The Black mother finding peace in the wider-than-human world. For Taylor, the story of a living, thriving gay man born and raised in a place like North Dakota. Not *Brokeback Mountain* and those tortured, closeted cowboys. Not Matthew Shepard, also tortured, struggling alone while tied to a Wyoming fence.

"How can the most biodiverse region create such a monochromatic way of thinking?" Taylor asked.

Shepard died in Fort Collins one mid-October morning in 1998, five days after first responders medevacked him to the hospital here. Mary lived on the flight path. She remembered the whir the helicopter made as it sped over her house with Shepard's brutalized body inside.

"In such a biodiverse place, our thinking is so limited." Taylor caught my eyes again through the screen. "Nature thrives on biodiversity. You know that."

• • •

My favorite Thomas Nuttall story happened in 1811. Only twenty-five and recently arrived from Yorkshire, England, Nuttall embarked on an exploration of the Missouri River, near the Dakota Territories—that immensely biodiverse region Taylor described to me. In his book on garden plants, David Stuart wrote that Nuttall "set out alone and got terribly lost." He had a habit, according to one contemporary, of engaging himself with the collection of plants "to the total disregard of his own personal safety and sometimes to the inconvenience of the party he accompanie[d]."

When Nuttall didn't return to camp in a timely fashion, his companions hired a group of Mandan or Arikara men to search for him.

The Arikara call themselves Sahnish: "the original people from whom all other tribes sprang." The Mandan also have names for themselves, including Numakiki ("people"), but Nuttall's companions did not record these names in this story. Instead, his contemporaries described the "Indians," who saw Nuttall carrying a gun. These men maintained their distance while tracking the Englishman. Afraid of who he saw following him, Nuttall sped away, "ducking into ravines and hiding in the brush." Despite his fear, Nuttall sometimes stopped to dig up specimens—roots and all—with the butt of his gun.

"My botanical acquisitions in the prairies proved so interesting as almost to make me forget my situation, cast away amidst the refuse of society without money, unprovided with every means of subsistence," Nuttall wrote. Except, he had every means of subsistence at his disposal. Had he engaged with the men who pursued him,

Nuttall might have learned a lot about the plants he collected. The 1998 version of Daniel E. Moerman's reference book *Native American Ethnobotany* catalogs two pages of uses for the common sunflower, *Helianthus annuus*. Moerman includes descriptions of how Mandan people powdered and boiled sunflower seeds to form cakes for portable food. But the twentieth-century naturalist's heavily cross-referenced 927-page volume required the invention and utilization of computerized database technologies unthinkable to Nuttall. It also required that Moerman acknowledge and record Indigenous wisdom and customs—something Nuttall did not do.

In 1811, Nuttall and the Sahnish or Numakiki men moved steadily toward the trading post without speaking to one another.

My daughter and I are from the American West. We love this place. But we no more belong here than Gene Autry, who sang so convincingly about a home on the range while misnaming bison, one of the key drivers of this landscape's current shape. Bison live here (*Bison bison*), not buffalo. Cape buffalo (*Syncerus caffer*)—with their swooping helmetlike horns, quite different from bison's shorter, clearly separate, spiked horns—live on the African continent. And water buffalo (*Bubalus bubalis*) live in South Asia. As with the Indigenous American people labeled *Indians* because the Europeans who encountered this hemisphere thought they had landed in Asia, calling bison *buffalo* reveals America's often-brutal roots in misguided trade. The word *buffalo* comes from the French word for an animal that produces beef: *boeuf*. So violent, to name a being not for how it lives but for how we consume it.

The roaming antelope Autry sang about also don't exist here. Though they resemble gazelles and elands of the African continent's

Antilopinae subfamily, the animals we call antelopes are actually pronghorns (*Antilocapra americana*), a species found exclusively in North America. Pronghorns have lived on this continent, in the West, since the Early Pleistocene. They can run faster than sixty miles an hour, much faster than any living predator found in North America today. Scientists believe *Antilocapra americana* developed their astonishing speed to outrun the American cheetah (of the extinct genus *Miracinonyx*) who roamed this continent before the Holocene. The scholar John A. Byers says we see in pronghorns' modern-day behavior "the ghost of predators past."

"Pronghorn were formerly very abundant throughout their range. Around 1800, more than 40 million animals were present, with about 2 million in Colorado," write David M. Armstrong, James P. Fitzgerald, and Carron A. Meaney, authors of the second edition of *Mammals of Colorado*, published in 2010. "By the early 1900s," the authors report, "there were few pronghorn remaining" due almost entirely to uncontrolled human predation and pressures put on these free-roaming animals by the invention and distribution of barbed wire.

The state of Colorado prohibited pronghorn hunts between 1893 and 1945, then allowed hunting with restrictions and management practices that prioritized the long-term viability of the herd. And more ranchers began using pronghorn-friendly fencing materials that left space at the bottom for the animals, who are not leapers, to crawl underneath. Thanks to these measures, eighty-five thousand pronghorn roamed Colorado in 2019. The continent's largest population lives to the north, in Wyoming, where wildlife management teams counted close to 400,000 heads in 2019.

Bison roamed a larger original range than pronghorn. Historians think more than sixty million bison lived in North America at one time. From the Allegheny River basin—in present-day West Virginia, Pennsylvania, and Ohio—to the Colorado River, huge herds roughed open the earth as they walked across the West, creating wallows, in which grew wildflowers, which drew insects and birds and small mammals, who in turn drew predators. When it rained or snowed in the high desert country, the wallows' dips and cups soaked up water with that now-active and aerated soil, and the hydrated ground allowed more living beings to thrive—prairie dogs and burrowing owls and rattlesnakes and black-footed ferrets and mound-builder ants and grasshopper mice and grasshoppers and mountain plovers and thirteen-lined ground squirrels and all the herbs and shrubs and flowers and grasses of the long and tall grass prairies that those many many bison roamed.

But even by the time Nuttall walked the interior digging up plants with the butt of his gun, Euro-American settlers had severely contracted the bison's range. By 1872—when Dr. Brewster M. Higley, lately of Kansas, wrote the lyrics that Autry helped make famous in 1947; or maybe Bing Crosby sang it into American hearts in 1933; or maybe it was the Black saloonkeeper whom John Lomax said sang it to him in 1910; it's no matter when, already then—white men willfully, insatiably, even genocidally slaughtered almost every roaming bison in America, to make way for the sedentary culture in which I take part.

I can't dig in my garden, my two-tenths of an acre of some homesteader's one hundred and sixty, without digging up all this old dirt.

Manifest Destiny. Settler colonialism. Complications and complicity. Six all-Black cavalry and infantry regiments, known now as the buffalo soldiers, advanced opportunities for many Black people, and prepared men to serve as rangers and conservators of many of America's new national parks. But also, as the great-grandchild of

Tenth Cavalry sergeant Henry Parker writes, these men served in the segregated United States Army as "a part of the Plains Indians' nightmare." Despair. Disease. Displacement and grief. Colonel John Chivington commanded the slaughter of nearly 230 Cheyenne and Arapahoe people, mostly women, children, and the elderly, during an eight-hour November 1864 assault that historians call the Sand Creek Massacre. Blood and pain. Removal and renaming. Despite 2020's historical reckonings, developers of a newly built neighborhood cut into a swath of tall grass and wild alfalfa a few miles from my garden and poured asphalt on a road named after Chivington, honoring that brutal man's legacy still.

If, from the window in my study, I look around the blue spruce to one of our neighbor's yards, I see a mostly dead cottonwood tree. The tree stands seventy feet tall. And though she still grows some leaves in spring, sending puffs of snowy fluff around the neighborhood each May, all but two of her eight prominent limbs are gray and bare all the year. Through those limbs, I can watch the sunset. As if the fading red light wants nothing more than to be caught up awhile and held before disappearing.

Before this subdivision, before all these lawns, before landscapers put in aspens, which prefer thickly clustered groves that lawn mowers and garden pruners prevent, before human-sited junipers and sidewalks—grew rabbitbrush and blue grama and fringed sage and milkweed. Near the tributary for which this neighborhood is named grew the West's largest deciduous tree. The cottonwood. Sturdy and steady, even past the prime of her full show of yellow-green flowers.

Five miles east as the crow flies, and close to the interstate that could take us straight up toward Fort Laramie in two and a half hours, there once stood a cottonwood people here called Council

Tree. That tree lived perhaps 120 years. Even after death, the tree kept council another seventy. A local Arapahoe leader met white settlers under the tree to resolve disputes and barter agreements. The Arapahoe also held council there among themselves. As did the Ute. Even after the forced removal of the Ute and Arapahoe, whose people lived in this region more than one hundred centuries, the tree stood in that place until a fire in the 1950s. In 2008, developers built a shopping center nearby, with a Target and a Sprouts and a library and restaurants and a Lowe's and a shoe store and a Sephora and a Staples. The road we use to get there: Council Tree.

"History will be kind to me," Winston Churchill said, "for I intend to write it."

I didn't see the story of the Sand Creek Massacre in school history lessons Callie and I worked on at home in 2020. I should not protect her from that story the way I protected her from the dead bunny. History is perennial, returning regardless of how poorly it's tended. The world is full of suffering. Callie should learn about the removals and diminishments and murders that opened the pathway to the settlement of Colorado as we know it. We should live honestly inside our history.

When Callie was in third grade, I chaperoned a field trip to see some of the homes of the area's original settlers, including that of Elizabeth Stone. Rather than the oxen many homesteaders used, Stone chose dairy cows to pull her Overland Trail wagon. She milked them every morning. By evening, thanks to the westward road's churning, Elizabeth Stone had sweet butter to sell to hungry émigrés. Those cows, and the hospitality they allowed Auntie Stone to extend, made her a respected and powerful woman. Her name's all over town: Elizabeth Street, Auntie Stone Street, Auntie Stone's Kitchen. The Elizabeth is Fort Collins's fanciest hotel.

Much rougher than Auntie Stone's—dirt floor, no second story—another cabin sits in the park. This belonged to Joseph Antoine Janis. In 1858, Janis came down from Fort Laramie, where he worked as a scout, fur trader, and interpreter. He settled along the Poudre River and founded the town we now call LaPorte. Years before, in 1836, just two years after Nuttall trekked through Fort Laramie, Janis had already walked this far south. He remembered the stretch of river as "the loveliest spot on earth" and staked his claim in 1844, settling there as soon as the Nebraska Territory opened to homesteading. He beat the gold prospecting boom by a year.

Our guide described Janis—born in Missouri, the son of a French trapper—as also a French trapper. It takes some people countless generations—and what blood?—to be thought to belong here. Janis married First Elk Woman, a member of the Oglala Sioux nation, and together they had several children. In 1877, a US government order required all Lakota people in the region to move to Pine Ridge Reservation. The tour guide asked Callie and her classmates to imagine having to decide whether to keep everything they owned and stay in a home they built with their own hands, or leave it all behind and go live on a reservation.

With his family! I wanted to interject. Janis didn't have to choose whether to stay with everything he cared about or go to a reservation. His choice was whether to remain with his family or divorce them, deny them, abandon them, for the sake of property and wealth. He had to choose whether to play a direct role in the campaign to destroy and diminish the lives of his own children. The guide's words made it clear that no one extended First Elk Women or their children the right to exercise even such a grotesque choice.

"He chose to leave with them," our guide told the students. "By doing so, he stopped being a white man and became one of the Sioux." She said this as if Janis made a surprising decision. As if she considered his decision a shame.

I hope my neighbor never cuts down the cottonwood tree. The ravens, the finches, the doves, and the nuthatches sit on those high, leafless branches, surveying their world. May they keep at least that.

In his essay "My Adventures as a Social Poet," Langston Hughes writes:

> Poets who write mostly about love, roses and moonlight, sunsets and snow, must lead a very quiet life. Seldom, I imagine, does their poetry get them into difficulties. Beauty and lyricism are really related to another world, to ivory towers, to your head in the clouds, feet floating off the earth.

I love these Colorado clouds. Like cotton balls some days. Gone the next. Some days gathered near the Front Range peaks so I can't tell where the mountains end and clouds begin. One thing always leads into another here.

The westward expansion had been in full effect for decades before Colorado became a state. Prospectors found gold here. The Guggenheims found silver. Seeking wealth in the region, more and more people from elsewhere arrived. Their arrival introduced more and more broken and annulled treaties—altering and imperiling the lives of more and more of the region's original inhabitants. The size of the territory encompassing what would become Colorado shrank and shrank, each time changing names—Louisiana Territory,

Jefferson Territory, Wyoming Territory, Kansas Territory, Colorado Territory. New states squared off around it, officially adding their populations and land masses to the United States.

I like to think of my birth state as expansive and welcoming. When it was finally ratified in 1876, lawmen wrote the state's constitution in English, Spanish, and German, offering welcome to many men of European descent. But one of the reasons it took so long for Colorado to earn its statehood is that the territorial government repeatedly presented articles of constitution to Congress that withheld from Black men, "mulatto" men, and Native men the right to vote or serve on a jury. Even after the nation ratified the Thirteenth and Fourteenth Amendments granting the full rights of citizenship to any man born or naturalized in the United States, *three times* leaders of the Colorado Territory tried to create a state where men like my husband, my father, Langston Hughes, could serve on no jury, could cast no vote.

I am not always sure that I belong in Colorado. Though I have my little plot of land here that I love, I am nothing but a settler in this state. And not always a welcomed one.

I try to reinstate some of the life that the land once knew. Its "roses and moonlight, sunsets and snow." In our sunflowers on a mid-August morning, as many as ten goldfinches perch, nibbling on petals and pistils and seeds. They blend into the foliage. I watch for the sun to catch the rustle of their golden feathers.

Sunflowers are native to the American West. They grow wild here. Nuttall identified both *Helianthus petiolaris* and *Helianthus nuttallii* on his westward treks. Indigenous people have cultivated sunflowers on this landscape for four thousand years. Thanks to Van Gogh's famous paintings and to the mammoth sunflowers of Russia, whose genetic alterations circled back to this continent to help grow the thirteen-foot-tall, hybridized beauties in my yard, I thought of Europe when I first saw them. But starting with the Spanish and the French and continuing through to Nuttall and his Anglo- and Euro-American fellows in the nineteenth century, people pulled sunflowers—like echinacea and black-eyed Susan and coreopsis—from the land here. Shipping them to gardens in Europe and England.

Bred to have only one large head and concentrate growing energy for a yield of larger seeds, Russian sunflowers differ from the Native American versions whose genes remain strong in my yard. The sunflowers I grow boast many shoots and branches, from which small heads of sunshine bud and blossom from July until the first hard frost. I can't blame anyone for wanting to gather and cultivate in their own gardens specimens from across the American West. If I ever have to go away from here, I'll be loath to leave these flowers.

I was born in Colorado. I think of this as the place I belong. But this is not true. The cottontail rabbits who live and sometimes die in our yard—they are the natives. Though they bear the name of a white man from England. Some of the plants they like best—prairie mallow, wine-cups, little bluestem—are natives here too. I hate to think of all who have died here so I could claim this as my home. Some I had a hand in killing, and some deaths, though not by my own hand, I benefit from.

I hated stumbling over that dead rabbit. Hated worrying how many will die from this new disease. I want to set a path toward survival. Despite everything, I want my family, so many of us, to lead a very quiet life.

Nuttall's larkspur in the rock garden

Metaphor of America as this homegrown painted lady chrysalis

My head has come off
and by a string of my own creation
        is dragged what remains
of my last meal. Here, too, you see
my waste, and my brothers' and sisters'.
        You can take this literally or not.
Whatever I might have been has dissolved.
        When you moved me, I shook
        like a leaf preparing for autumn.
The child panicked. But soon, I returned
        to my patience. Call it potential
if you're feeling optimistic. There will be wings.
Bright, brown, black. With just a little
        white to set things off.

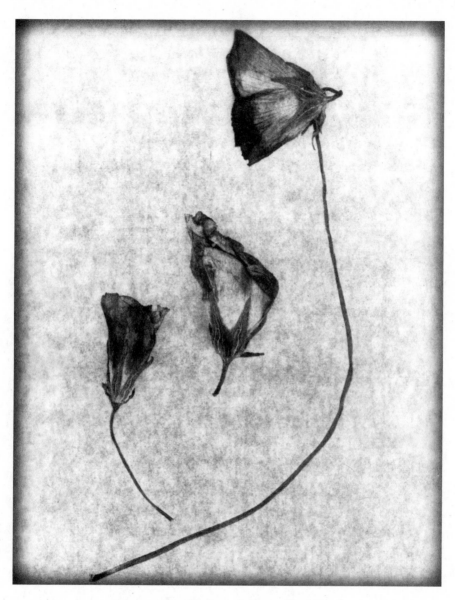

Wine-cups picked just before a hard frost

From my bedroom window, I noticed the first yellow leaves on the honey locust that droops over the back neighbors' fence into our yard. That was late August. I wasn't ready for fall. Much less for winter.

Sometimes, in the middle of July, I look out the window and think, *Soon this will all be covered in snow.* On a walk to the nearby pond where the Odonata gather to flare their lovely wings, I think, *In a few months this will all be frozen over.* Like looking at a locked door and thinking, *Someday this will fail us.* They frighten me, these thoughts of long months when I don't have my garden to give me something to do with my hope and my hands.

When Callie heard about how police with a no-knock warrant broke into Breonna Taylor's house and shot Breonna to death while she slept, my daughter asked if such violence could happen to us.

I tried to say no.

I had to say maybe.

Callie suffered panic attacks all through the summer of 2020, then into the fall.

Focused on something perfectly normal—completing her science homework, folding into dance stretches, unloading the dishwasher, playing in what she called her "imagination factory"—she'd suddenly start hyperventilating. She worried police would kill her.

When we drove past a police car, her decade-old body tensed until the threat passed. No one in our car had done anything illegal. But she knew the meaninglessness of innocence. She heard the news. Breonna Taylor. Sandra Bland. George Floyd, in Minnesota, under an officer's knee for nine minutes and twenty-nine seconds. Tyree Davis. Kwame Jones. Donnie Sanders. Michael Dean, in Texas, unarmed but shot mere seconds into a routine traffic stop. And, resurfaced from August 2019, another story from Aurora, Colorado, sixty-five miles south, where twenty-three-year-old Elijah McClain left a convenience store after buying iced tea. He used to go to the animal shelter and play violin for the cats and the kittens.

We lived in a season of fear so constant, so common, it almost seemed reasonable.

The locust leaves began to change. A yellowing here. A reddening there. A crisp, dry wind that sent me back inside some mornings.

Leaves dropped. Soon most everything around us looked to be dying or dead.

In fall 2020, I expanded the oval in the center of the lawn to enlarge the flower-growing capacity of that first reclaimed bed. After several springs spent weeding the persistent rhizomes and stolons of bluegrass and crabgrass and fescue from newly established flower beds, I learned that burying sod in layers of cardboard, compost, newspaper, topsoil, and mulch does not stop grass from growing back and threatening to choke our flowers. I couldn't just cover what I didn't want. I had to completely remove the water-thirsty sod.

With an edge trimmer, I cut an outline for the new patch. Then I worked the oval-tipped shovel under the space the edger offered, pushing the metal below the grass roots for leverage. Inches of sod pulled away from the dirt below with a sound like separating Velcro. I repeated the process until a large swath of sod sat on the surface like a poorly secured toupee. Then I got on my hands and knees. No garden gloves—I needed the dexterity my bare hands offered to roll the heavy carpet of sod and heave it onto a wheelbarrow. Into the space where the lawn had been, I tilled loamy-smelling compost from our backyard bins.

A little clover grew in the old sod, some dandelions, but it was healthy. It seemed a shame to throw the thick rolls into compost. I used some to patch a dead spot in our own lawn, and the neighbor with the cottonwood agreed to take the rest for his yard. I sweated as I pushed the grass-weighted wheelbarrow on four trips across the street. Each roll heavy as ten-year-old Callie, whose sleeping body I could hardly carry anymore.

Via a recent shipment from an online nursery, I had two *Solidago glutinosa* starts, a species Thomas Nuttall identified and named thanks to the Wyeth expedition. The nursery's catalog promised golden sprays of flowers from wand-shaped, short-branched arrays. After a spontaneous purchase at the local grocery store, I also had a gift-wrapped rust-red and gold chrysanthemum. To give them space to grow into their full twenty-four inches, I sited the new plants carefully in the expanded eight-by-five-foot section at the center of the yard. I also spread a handful of tick-size purple prairie clover seeds, Colorado bee plant seeds, and seeds from spiky echinacea pods the size of the sweet gum balls that gathered around trees the autumns I lived in North Carolina and Virginia.

Finally, I planted thirty-six orb-shaped bulbs: little beauty wildflower tulips, Colorado proud triumph tulips, the poet's daffodil. Soon the new patch looked like an island of soil and months-from-now dreams. White, yellow, orange, reddish-pink, and ivory-edged black cherry blooms wouldn't manifest for eight months—nearly as long as I waited to meet Callie. Like the labor it took to bring my daughter into the world, many ministrations in this garden happen while I'm on my hands and knees. I am often supplicant, begging the earth to provide what I need.

The September weekend I buried the tulip and daffodil bulbs, a wildfire grew to 124,000 acres right around the place where we scattered Mary's ashes and the children tied red ribbons on the sage.

The Cameron Peak Fire began in mid-August near Chambers Lake, sixty-five miles west of our house. By the weekend of September 27, the fire grew into the third largest in Colorado history. Officials suspected human activity caused the blaze.

We experienced above-average temperatures that September, as high as 91 degrees. Winds and dry conditions encouraged the Cameron Peak Fire to spread toward populated communities, affecting residents of more than one thousand homes. I tracked fire maps as more and more places I knew fell into voluntary evacuation, then mandatory evacuation, and then, sometimes, active fire zones.

I almost didn't plant the tulip and daffodil bulbs that September Sunday. The cloth masks we wore for COVID did little to protect us from toxic particles in smoke. I found breathing difficult. My eyes stung and felt scratchy. Tears welled to prevent further damage to my lenses.

Back in 2018, while the Mendocino Complex and Carr and Ferguson Fires burned not far from where we stayed, Ray and Callie and I spent several weeks in Northern California. A total of 1,975,086 acres burned in 7,948 California fires that season. The California Department of Forestry and Fire Protection (Cal Fire) reported 100 confirmed fatalities and 24,226 structures damaged or destroyed during the long fire season of 2018, which was, at the time, the most devasting in California history. All that smoke scarred my eyes, and my vision has never been the same. Being that close to catastrophe changed how I see.

Living in this part of America, I know that everything I love is liable to go up in flame.

One of the last things Elijah McClain ever said was, "You guys started to arrest me, and I was stopping my music to listen."

"If you keep messing around, I'm going to bring my dog out," I can hear an arresting officer on the bodycam video say. "He's going to bite you. You understand?" In the video, McClain tries to roll over to vomit. The officer's knee on his torso made him sick.

In the transcript, McClain cries and he begs and he wonders. "That's my house. I was just going home," he says. "I have no gun. I don't do that stuff. I don't do any fighting. Why are you attacking me? I don't even kill flies!"

McClain suffered from a blood circulation condition. When he went out into the harsh dry Colorado air, he often wore a mask. In pictures, his smiling face has the same warm-brown shade and softness as Callie's. The person who called 911 to report McClain said he looked "sketchy." The caller said Elijah McClain didn't belong where he was looking the way that he did.

Early in the bodycam video, I hear an arresting officer demand that McClain "stop tensing up." Things escalate quickly. It's hard to make sense of what happened from the transcript. One of the officers asks, "When did the fire start? People come in there just to start fires." Elijah McClain is facedown on the grass at this point.

No. At this point Elijah McClain is long dead.

Late in the bodycam video, one of the arresting officers says, "He was put into a carotid, and he didn't lose consciousness." Even though McClain weighed only 140 pounds at five foot six inches—that's how tall I am. That's four inches shorter and sixty pounds lighter than the average white or Black American man—an officer says, "He has incredible strength."

"Stop, please!" I can hear McClain begging.

In a break from late September's wildfire smoke, I planted three dozen small brown tulip and daffodil bulbs and could not stop thinking of Elijah McClain.

At one point, the arresting officers told McClain to lie down in the grass.

"We're still struggling with him," one officer says. "I have fire on the scene." At this point in the transcript, I understand that someone called the fire department for support.

"Please help me!" McClain cries.

In an ambulance, a paramedic from the fire department injects McClain with five hundred milligrams of ketamine. Over 30 percent more than the dose required for a man of his size.

In the ambulance, McClain goes into cardiac arrest. He never recovers.

What does any of this have to do with nature?

It's the dogs. The flies. Those police officers and EMTs treated McClain like a dog—an animal. Like a horse. Like a pest. Like a fly.

It's that Elijah's mother had to bury him.

Not into my garden. Not into the comfort of my family. Not kneeling in the grass. Not on a walk, on a drive. Not even in my own bed. I've got nowhere safe to go in America.

Callie knows this already.

Death spins around us. Like smoke.

In *Leaves of Grass*, Walt Whitman wrote, "Every atom belonging to me as good belongs to you." I believe that must be true. His energy ghosts my garden: Elijah McClain.

Burying those little brown bulbs that fire-filled September, I thought of the day I became a Black child's mother.

The garden kept its own council. It waited. Not on me, but on warmth and lengthening light.

Gardening books suggest I should keep some plants in the yard for winter interest. Variety to gaze upon during the bleakest months. I left the long stems and many tiny round pods of blue flax. Dry by then, and tan as a Thanksgiving wheat display. The shin- and knee-high Nuttall's larkspur, whose inch-long, three-chambered, peapod-shape seed sacks turned a tannish brown in July and August. They split from the top to spill tiny black pellets, which I directly sowed where I wanted them or let the wind scatter for a new crop in spring. I let stand the feathery plumes of blue grama, whose name must come from the purplish-blue hue its stems and panicles showed off in fall and winter. Dragon heads of dried hyssop. Seed heads of scorpion weed too. When most color fled the region for the season, these bright brown stands of shape and contour gave an eye delight.

Even the evergreens dimmed for the long cold season. The juniper and blue spruce turned plumlike gray as they slowed chlorophyll production and the waxy protective coating on their needles showed. The wintercreeper turned a duller shade of green, concentrating on storing energy for spring growth.

Some bees and butterflies who visit the garden lay eggs inside the hollow reeds of desiccated flower stalks. Under mats of fallen leaves, a few caterpillar larvae overwinter. Working in the garden, I acquire the patience to believe in the value of so many lives—all those different compositions of atoms I may never see but want to support.

After catching glimpses of a blue jay who nested in a stand of lilac and leafy shrubs in one of our back neighbors' yards—blue on the wing as the bird dashed in and out of cover—we put out some of the birds' favorite feed in August 2020. Soon, we saw more blue jays. One day, as fall came down on us, five at once descended, grabbing nuts for their caches.

The jays called to each other when I entered the yard. Caws I learned to differentiate. Some anticipatory, some excited, some commanding, some disappointed in the lack of new seed. Several birds cooperated with each other. Others competed. I spent days, several minutes at a time, watching and listening to the jays who started coming regularly to sit on the locust and aspen and hawthorn and maple and fence posts and feeders in and around our yard.

After Thomas Nuttall trekked from Saint Louis to the mouth of the Columbia River, passing through modern-day Nebraska, Wyoming, and Idaho, and after he sailed down the West Coast, disembarking occasionally to describe the flora and fauna of Oregon and California, he took a trip to Hawaii, known then as the Sandwich Islands. There, he named a subspecies of the tree snail *Achatinella apexfulva* and collected *Schiedea nuttallii*, a flowering plant endemic to Oahu though currently registered as endangered due to introduced threats similar to those faced by the tree snail. When

Nuttall's sojourn ended, he sailed around Cape Horn and back to Boston with a collection of hundreds of new species of plants, mollusks, and crustaceans to be described for English-speaking eyes and ears. Thanks to Nuttall, this nation's wealth and value became more calculable, and extractable.

Saying he had done such a service to "mankind" it would be folly to charge him, the owner of the shipping company refused Nuttall's money when he walked into the Boston office to pay his fare. Thomas Nuttall played a role in the identification and naming of close to 6 percent of all North American plants, and some animals too. Nuttall's larkspur (*Delphinium nuttallianum*), Nuttall's saltbush, Nuttall's oak, the Pacific dogwood (*Cornus nuttallii*), sego lily (*Calochortus nuttallii*), alpine golden wild buckwheat, northern mule's-ear, the yellow-billed magpie (*Pica nuttallii*), and so many more. Close to October, the Nuttall's cottontails in our yard prepared their pelts to blend into the coming long gray winter.

Prompted by a recognition that Euro-American men's legacies dominate this landscape, in August 2020 the American Ornithological Society (AOS) voted to change one bird's name. *McCown's longspur* became the *thick-billed longspur*, no longer using the little brown bird to honor a Confederate general who led troops in wars against Indigenous American nations. "With the nation as a whole reckoning with the structural racism embedded in U.S. society," wrote Jessica Leber, senior editor at *Audubon* magazine, the AOS, which confers official names on birds, chose to acknowledge "pressure from a vocal group of birders who want to see sweeping changes to bird names, as well as to the process for determining them."

Initiatives like Bird Names for Birds pushed to rename around 150 birds who bear honorific or eponymous titles that also celebrate violent histories. Townsend's solitaire, a gray bird with giant eyes, is named for a nineteenth-century naturalist known to have robbed Native American graves to send human remains east for scientific scrutiny. Based in part on skulls Townsend provided, *Crania Americana* classified Native Americans as a separate species from white people. Samuel George Morton, the author of this phrenology book, also received skulls that John James Audubon decapitated from the bodies of fallen Mexican soldiers during the Texas border war in 1836. Bird Names for Birds wanted a new name for Audubon's oriole, the gloriously bright, yellow-bodied bird with a black hood, black wings, and a black tail. The same goes for Bachman's sparrow, a little brown job named for a man who supported the continuation of the institution of slavery. In an 1850 book expounding his ideas about the superiority of the white race's beauty and intelligence, the Reverend John Bachman compared people of African descent to wild boars and "the dull cart horse," claiming that "in intellectual powers the African is an inferior variety of our species."

Bird names that honor such men "are a memorial both to the colonial system that wove the fabric of systemic racism through every aspect of our lives—including the birds we see every day," wrote two Bird Names for Birds activists in an August 2020 *Washington Post* op-ed, "and to the individuals who intentionally and directly perpetuated that system." There's no escaping the legacies of this nation's violent, exclusionary practices. But maybe they don't have to live on in the names we use for the vibrantly colorful lives around us.

The taking and staking and claiming and dividing and naming of the American West had everything to do with who had the money and cultural capital to call it their own. Morton paid Townsend $50 (equal to nearly $1,500 in 2020) for two Indigenous people's skulls. Wyeth paid the nineteen men on his expedition $250 each for their eight-month engagement, a payroll obligation equivalent to $143,000 in 2020. Whether your own or someone else's, it took a great deal of capital to explore and describe the world.

Returning from the Wyeth expedition, Nuttall found himself nearly destitute. Perhaps fortunately, soon after, his wealthy uncle Jonas died, leaving Thomas his estate. Poor man. The inheritance stipulated that Nuttall live nine months of every year in England. In those days, it took the better part of a month to sail from England to the East Coast of America. Accounting for his return trip, Nuttall would have only a month in the United States. He couldn't manage the long journey to his beloved American West during such a short time. One year, Nuttall tried to circumvent the residency clause by leaving England in September and returning the next year in March. But this put him in America when most of the flora are dormant or dead.

I wonder why Nuttall's uncle added this clause. Having built his business during the former colony's insurrections—the Revolutionary War and the War of 1812—perhaps the senior Nuttall didn't appreciate his nephew's love for such a traitorous state. Or maybe he worried for the younger man's safety. Long treks through American swamps and deserts and mountains and riverbeds wrought countless ills on Thomas Nuttall's body. Maybe his uncle exercised his control out of a desire to take care of a young man he loved. Or perhaps jealousy inspired Jonas to rein in his footloose nephew. Maybe he thought exploration an unsuitable occupation for the heir to a self-made printing fortune and he wanted someone to watch over the great house he'd built outside Liverpool—Nutgrove Hall. The reason is less important than the outcome. Thomas Nuttall's 1842 acceptance of the proceeds from his uncle's estate marked the end of his time as an American explorer, botanist, namesake, namer, and scribe.

By October, most of the Nuttall's larkspur in our yard looked like their summer selves but browned and stiffened—as in legends where spells transform every living being into something like stone.

The Cameron Peak Fire exploded to 164,140 acres by the morning of October 14, leapfrogging the 139,007-acre Pine Gulch Fire that had held the record as the largest fire in Colorado for fewer than three 2020 months. Mandatory evacuation zones spread to twelve miles from our house. Smoke billowed over us, darkening the sky to a rusty orange. We kept the lights on inside the house all day.

I checked our survival kit: flares, flashlights, blankets, spare clothes, portable tools, emergency meal rations, first aid supplies, water filtration tablets, important documents, running shoes, seeds, a good knife. For the first time since the shutdowns started in March, I packed a suitcase.

When we still lived in the earthquake-prone Bay Area, I added a baby sling to our go bag a week after I delivered Callie. If we had to run, I could carry the baby and keep my hands free.

Sometimes I think I worry about catastrophe too much.

Sometimes I think I don't worry enough.

Visiting our Oakland apartment late in my third trimester, our doula noticed all the heels I kept near the door. "I never wear high-heeled shoes," she said.

Registering my curious expression, her tone changed from incredulous to informative. "I grew up here. My parents were Panthers." This was shorthand I understood. In Oakland, California, some people revered and relied on the Black Panthers for the ways they demanded racial and economic justice. Others maligned and attacked them for the same reasons. "My family taught me I need to keep both feet firmly on the ground," she said. "You never know when you might need to run."

As the fire spread, I placed our go bags within easy reach.

Ray taught his classes online, as he did every Monday, Wednesday, and Friday. Callie kept up with her online schoolwork. A normal day. With heavy smoke and ambient panic. Friends who lived west of town packed what they could and evacuated. They left most everything behind.

Ray and I touched base with a friend to whose house we would flee, talking over how to account for the complications of sheltering with someone else's family during a pandemic. We made sure my parents also had an evacuation plan in place. At six p.m., after the worst of the smoke, I watched eight pine siskins gorge at a Nyjer feeder. The little brown speckled birds had sheltered in the blue spruce, not eating during the smokiest hours. None of us is immune from the chaos of fire.

On one of those San Francisco days that made me think I lived in heaven—temperatures in the high 70s, the sun softly filtered, even in the fog belt—I sat at a sidewalk table of a Castro neighborhood staple called La Mediterranée, eating crepes and salad with a colleague. A gentle breeze blew, scented with citrus blossoms, ocean salt, and the bright, sweet leaves of mariposa shrubs. People pay good money to get such scents in candles and air fresheners. Out west, they waft through the air for free.

But nothing is free.

"It's earthquake weather today," said my colleague.

"Maybe so," I agreed.

Earthquakes have shaken me awake many times. My mother

once watched a glass bookshelf totter toward her while she sat on the couch talking on the phone with her brother. Twenty-three-year-old Anamafi Moala died when the 1989 Loma Prieta earthquake collapsed the section of the San Francisco–Oakland Bay Bridge on which she drove, though hundreds of other cars safely crossed the bridge in the moments before the tremor. Earthquake weather is weather that reminds me my hold on paradise is brief.

I called Ray the moment I got back to my apartment that beautiful day. "She called it earthquake weather," I said.

Ray did not laugh.

"I know there's no such thing, but I got worried."

Ray stayed quiet.

The epicenter of one of the costliest wildfires in our nation's history was a town in Butte County, California, called Paradise. At 6:30 a.m. on a partially sunny 50 degree morning in November 2018, someone reported smoke.

As US Forest Service historian Lincoln Bramwell writes, "The Camp Fire (named after a road near its origin) moved over 10 miles in four hours after it started." Assisted by years of drought, drying heat, and katabatic winds that gain speed as they blow in from the east off the desert, the blaze swept downhill toward the twenty-seven thousand residents of Paradise.

The flames moved, according to Bramwell and other sources, at a rate of one football field per second. Within a few hours, the fire had razed a significant portion of the town. It burned about seventy thousand acres and contributed to most of the eighty-five total fatalities on that first day. For the next seventeen days, until crews contained the blaze, the Camp Fire burned 153,336 acres and destroyed 18,804 structures, including 14,000 homes. Assessments suggest the Camp Fire caused upward of $16.5 billion in losses. No one goes to Paradise without witnessing disaster.

At the time I called Ray about the weather, I was thirty-four years old and had never considered what might happen to someone I dated should a natural disaster arise. Ray and I had known each other only a couple of months, but as we talked my voice rose to a panic. "What if something *did* happen? You live over in Oakland, and I am here in San Francisco. How would we find each other? What would we do?"

Ray finally spoke. "You get yourself to your godparents' house," he said with conviction and clarity. I could take several different routes to get there, some that required no bridge. "There are people who love you and will worry about you. They'll need to know you're okay. Stay with them and wait. I'll make my way to you."

That's when I knew that I loved him.

Ray's plan to find me and to keep me safe gave primary consideration to community.

The East Troublesome Fire—whose name sounded like the title to a classic country western song—wasn't even an issue on our side of the Continental Divide on Tuesday, October 20. But it hopped to the east side of the divide on October 21, burning more than 120,000 acres by sunrise on October 22. The fire became the second largest in Colorado history over the course of the day, eventually growing to about 194,000 total acres, scorching several communities, and killing one couple in their home.

To fight the East Troublesome Fire—and two others that threatened houses and land close to highly and expensively populated areas—fire management officials pulled some crews off the then-208,663-acre Cameron Peak Fire. Scholars like Lincoln Bramwell stress that, because there is so little divide between our built environments and western areas considered wilderness, we can expect more catastrophic "fires on the forest edge that involve homes." Bramwell's assessment in his book *Wilderburbs* offers more proof that the imagined divide between what we call the wilderness and what we claim as the domestic is false and catastrophically dangerous.

A bit south of the East Troublesome Fire's burn scar, crews completely contained the Lefthand Canyon Fire near the busy town of Boulder by the end of October, and improved weather held back the 10,105-acre CalWood Fire near Longmont, thirty miles from our house. But, fanned by winds as fast as seventy miles per hour, and fueled by dead trees and invasive grasses, the 176,878-acre Mullen Fire burned since September 17 in southern Wyoming. For a long while, the Mullen Fire threatened the Rob Roy Reservoir, the water source for Wyoming's state capital, Cheyenne, less than an hour up the road from Fort Collins. By late September, the Mullen Fire spilled over the state line into the northern reaches of our county.

When the Cameron Peak Fire flared bigger in October, I checked the calendar. Had it already started the day we drove up near Beaver Meadows to spread Mary's ashes? I could believe that. I could believe something that seemed a small, distant inconvenience could grow to consume my every waking thought.

Terror can be like that. Terror can be someone else's problem, somewhere else's problem, until it spills into my own backyard. Seeps like smoke into my house. How many friends did I speak to in 2020—that wild, deadly year—who told me, "I didn't know things were this bad until now." There should be a question mark on that sentence, but I am past questions. All the evidence has already, and always, been with us.

Living so close to the feet of the mountains, we're inside a convergence zone. Storms smash into the Rockies and pour out the last of their wrath. Hot air and cold collide, especially when seasons change. The resulting strong winds aided the Cameron Peak Fire the day it took over as the largest fire in the history of our state. It felt as if a giant shook our house like a toy. Like a giant hammered our windows, our roof, and our doors. I could only brace for the winds. I couldn't stop them. In the fire zone, eighty-miles-an-hour gusts blew flames south and east, out of designated wilderness areas and toward backyards. Crews of hundreds of firefighters built firebreaks, set back burns, dozed lines. Retardant tankers flew when winds allowed. Still, the fire gained an additional forty thousand acres in four days.

The Cameron Peak Fire was a perfect metaphor for many things: climate change, habitat encroachment, the American healthcare crisis made worse by a global pandemic, racialized injustice and brutality. For smoldering issues that flared up in those years, months, and days.

And, also, the fire was its very own thing.

Smoke-saturated wind rattled the house, yanked drainpipes, whipped the hollyhock stalks. Finches and siskins who lived in the blue spruce twittered loudly. They zipped past my window in their climb, dive, climb flights toward the sunflower patch. Then they rode the bobbing flowerheads and took their feasts.

What looked to be a beautiful cotton ball of a cumulus cloud moments before stretched wide and grayed. A pyrocumulus cloud. Part of the weather systems extremely large fires create when fire-heated air rises and expands, mixing with cooler air higher up. Such clouds forge their own lightning and produce fire-flaming winds.

By the end of the 2020 Colorado fire season, more than 600,000 acres burned within as close as ten miles from our house. That's nine hundred and thirty-eight square miles. I pictured flames burning New Orleans and Wichita and Durham and Minneapolis–Saint Paul and Sacramento and Boston and Buffalo and Newark all at once. That's how much land burned outside our door. A swath of America the size of the cities of Houston and San Diego combined. Ash rained. The air outside looked redder and redder. As if I had a bloody scrim over my eyes.

House finches flitted red-bellied bodies over the prairie project then settled on the edge of the birdbath. Cinders leached acid into the water. From the southern migratory path in New Mexico and Arizona came reports of thousands of birds dead on the ground. Falling out of the sky. Emaciated and destroyed by exhaustion, starvation, and smoke.

The fire was not burning the day we went to the meadow to spread Mary's ashes. It started four days later.

The fire had been burning for centuries when we went to the meadow to spread Mary's ashes. It will never go out.

Within days of our 2013 arrival in Colorado, Ray and I left Callie with Aunt Mary for two nights. Then we embarked on a university-sponsored bus tour on which new and newly promoted professors and administrators learned about work the land-grant university does around the state.

I forgot the name of the man we met in Cañon City during that August trip, and I could be conflating my memory of his dark brown hair and Tom Selleck mustache with my experiences of other men in his field, but I do clearly remember his smile. Pure joy I don't see often in men.

Speaking about a five-day fire that ripped through the area around Cañon City two months before, burning all but four of the fifty-two structures as well as large swaths of forested land at Royal Gorge Park, the man described a crew who would spread native wildflower seed across the burn site. To prevent massive erosion and protect water quality in the river, and also to suppress the growth of adventitious nonnative species, crews would spread mulch made from most of the burned trees. A few remaining snags would serve as wildlife roosts while the rest of the area's vegetation grew back. Another crew would plant hundreds of juniper saplings in the spring. Without such assistance, regeneration could take hundreds of years.

After fires, sometimes even before they're fully contained, fire management teams send in two kinds of crews—Burned Area Emergency Response (BAER) teams and Long-Term Restoration and Recovery (LTRR) crews—who help rebuild healthy landscapes. Seeding, reforesting, installing erosion controls, rebuilding wildlife-supporting fencing, mitigating the negative impact of invasive weeds: after a big fire, BAER and LTRR teams treat the forest like a garden, with the needs of a massive landscape in mind. The man we met in Cañon City led such a crew.

He spoke as if we ought to know about the fire that ravaged his community, but all we knew was what he told us that day.

Unless the story catches the right ears, news often stays local—lingering solely with people who claim a clear connection to the catastrophe. Elijah McClain died only an hour away from us, and it was a year before we heard. About seventy-five fires burned in Colorado, Wyoming, Idaho, Montana, Arizona, New Mexico, Oregon, and Washington State in late October 2020. While 4,304,379 acres burned in California that year. So many lives lost and endangered that I may never know a thing about.

But the man from Cañon City showed us the devastation there. He trusted that, if we saw, we would care.

My family's first winter was on its way. We had lived in a temperate climate so long that Ray and I worried about the great silencing that comes with snow. The man in Cañon City wasn't worried. He knew the slow release of moisture from winter snowfall would aid recovery efforts.

The August Ray and I met him near the burn scar, the man smiled. "It doesn't look like much now," he admitted. But come spring, he'd play a role in returning his small section of paradise to life.

Before each year's first heavy snow, I bring every container on our patios into the garage. Our car huddles on one side, and I cover the rest of the open space with potted plants. Most die anyway. Flowering plants that had been full of life all summer turn stiff as the dried floral arrangements in a craft store.

In the fall of 2019, we rearranged Callie's room, installing a double bed, a nightstand, a new bookshelf, a desk. Rather than a little girl's room with a twin bed and a small work surface, she now had a room to hold her through high school. Lovely, but lacking the benefit of drawers, the antique kitchen table that once served as her desk sat in the garage awaiting repurposing all November and December 2019 and on through the first half of 2020.

Stored in its pot under that table, one dahlia plant gained strength to come back another year. Near Mother's Day 2020, I saw a stalk stretch long like a tendril. I put the container outside in late May, and that dahlia sent out bright yellow flowers that attracted the only hummingbirds I saw in our backyard that summer.

The dahlia continued to bloom right up to the last week of October. A tight green fist of a new bud slowly releasing yellow petals. The flower called my attention away from the encroaching fires and back to our garden. I brought the dahlia's pot into the garage again, hoping the buried tuber, the source of that fierce flowering, would make it through another winter alive.

I also brought in a pot full of rosemary, and the lemongrass and lemon verbena I keep on the back patio for their natural mosquito-repelling properties. I knew most of the plants wouldn't survive, so I trimmed and dried aromatic leaves for cooking—a little summer brightness for our plates, even on the coldest days.

One morning late that October, I looked out our bedroom window and marveled at the fully golden honey locust. The same tree whose first tinge of gold filled me with unmitigated dread in August.

I harvested the last of our vegetable garden. Sliced several still-green tomatoes onto a homemade pizza. I stirred the last of the chard, kale, and basil into mushroom risotto. I tried to honor what I love in the ways I could.

Pine siskins and house finches harvested sunflower seeds from the plot near the driveway. Mourning doves walked around the concrete eating seeds that rained down from heavy sunflower heads. A black-capped chickadee swooped over dried purple hyssop heads, while the hyssops' root balls bided time belowground. In the center plot, more birds pecked at the sharp echinacea, whose petal-free crowns I thought I left standing too long. They'd grown unsightly, but the only days I had time to trim them were too filled with thick smoke for me to work outside. By waiting, I prevented myself from robbing the birds of natural, nourishing sustenance.

I collected milkweed's cotton-born seeds from pods that burst and spread their white fluff all over our yard. To encourage genetic diversity in the next crop, I picked a few from this plant, a few from another. I placed these into a paper bag to deliver to the friend who takes our river rocks.

It is more than hope my garden gives me. Examples of resilience keep me coming back to walk this path in gratitude and wonder.

Without resilience, what is hope but a passing fancy?

Faith is the belief in things not seen. Or it is the hope that what has not yet materialized might, someday, manifest. I am loosely quoting Hebrews 11:1, in which people hope for some future when a crucial promise might somehow be fulfilled.

One of the hallmarks of faith is to believe in a promise and—though the promise has yet to come to pass, and may never in my lifetime be fully fulfilled—to find a way to carry on. To discover and honor what *has* come to fruition.

I dig up a lot of awful history when I kneel in my garden. But, my god, a lot of beauty grows out of this soil as well.

All morning, all day, all through the night of the last Sunday of October, snow fell.

Snow stuck to our roofs and our yards, our bushes, and our trees. According to the multicolored measuring stick Callie and I assembled as a snow gauge, nearly fourteen inches fell. Twenty inches up near Beaver Meadows, where the Cameron Peak Fire raged.

Dead and down trees continued to smolder even with the new moisture, but the snow and cold gave crews a boost in their efforts to slow the blaze. Cooler air and added moisture mitigated much of the worst of the fire.

Crews didn't fully contain the Cameron Peak Fire until early December. It took even longer to completely extinguish it. We always live with the threat of new and worse fires. But the reprieve helped. We could breathe steadily again, and we welcomed the difference.

Snow swallowed ambient noise and quieted the yard.

For months, it felt as if my whole world burned. What a relief to see the sky spill snow and not ashes.

I trust that some of the root balls and seeds and bulbs and seedlings I have dug into our garden's soil will soak in the slowly released hydration delivered by snow. Soon, but not too soon, our plot will explode into glorious multicolored blossoms.

Colorado blue columbine

A Garden of Gratitude

Though we often think of them as individuals, most plants prefer to live in clusters—in groves and fields and prairies and gardens—communicating with others of their kind and also with mycelium, with birds, with rabbits, with bison and prairie dogs and insects and trees and other flowers. It is difficult to survive, much more difficult to thrive, without a community on which to depend. This, then, is my thanks to my vibrantly beautiful community.

From the start, I had my family around me. Ray Black and Callie Violet, thank you. Thank you. Thank you. There could never be enough words for how grateful I am to have you two in my life. For now, let me thank you for sharing so many meals with me. And for being willing to dig in the dirt with me too.

Thank you to my father and mother, Drs. Claibourne and Madgetta Dungy, and my sister, Dr. Kathryn Dungy. Your smiling faces were the first beauty I knew. Thank you to my grandparents, great-grandparents, aunts and uncles, cousins, and my beautiful sisters-in-law (I'm looking at you, Rayshana). And thank you also to my chosen kin: Mary Tesch Scobey, Sarah and John and their families. Vanessa Holden and Mariama Lockington. Ellie and Jim Manser. Janet and Andrew. Kimberly Wilson. Charlene Hall. All the Megans. Tammi. Whei Wong. Bayliss and Drew. And you, Terry, for being a brother to Ray. For all these years, you all have shown me not just who I am but, also, and importantly, who I want to become.

Lucy Anderton, for your careful reading and more, my gratitude extends across oceans. Like a purple martin or a western tanager, my heart will always fly from here to there for you.

Mary Cook, fact-checker and sharp-eyed reader extraordinaire. You have 100 percent of my heart. (See what I did there?)

Sumanth Prabhaker, I am grateful for your eyes and ears and vision. Talking over these pages with you when they were still in their budding days was such a pleasure. Thank you.

Dionne Lee, your art is amazing. Remember that day when we sat beside each other on the stage at Colorado State University, and I said I wanted some images in the book I was writing, and you asked, "What kind?" The images you created using cuttings from our garden and the photos I sent are exactly what I had in mind. Thank you!

And Mary Ellen Sanger, thank you for those days you came to our house with your camera to take pictures of the garden. And thank you for taking Callie under your wing and encouraging her as she took pictures (first with your camera and, after we all saw she had such a good eye, with her own). Thank you for sharing your art on these pages, and for sharing your art with my daughter, whose art is, therefore, on these pages as well.

To the writer friends who sent me messages and walked with me and talked with me all through the making of this book. Suzanne Roberts and Kathryn Miles. Sean Hill. Ada Limón, Major Jackson and Matthew Zapruder. Taylor Brorby. Erika Meitner, Paisley Rekdal, and Keetje Kuipers. Brenda Hillman. Tiffany Han. Nadia Colburn. Laura Pritchett. Jami Attenberg. Renata Golden. James Hoch.

Alice Wu. Leah Naomi Green. Tara Powell. Dr. Lehua Yim. Jane Satterfield, Penelope Pelizzon, and Shara McCallum. Angie Chuang. Valerie Miner, Vanessa Hua, Aimee Phan, and Toni Mirosevich. Drew Lanham. Lauren Alleyne. C. M. Burroughs. Zora. I am proud and grateful to live in a world where your words live too.

To Claire Boyles. Mary Ellen and Joseph. Gillian Bowser. Bob and Pam. Steve Comfort. Andy. Deanna Robertson. Amy E. Weldon. Kaveh Akbar. Tori Arthur. Lisa and Tom Chandler. Nicolas Weasley and Jess Turner. And anyone who has offered me seeds or starts or spilled their sweat and love into the soil of this garden, thank you.

To Tim. Thanks for looking at the crap piles around us and finding sweet reasons to laugh.

To Lincoln Bramwell, for talking about fire during those crazy COVID-19 days when our kids played together in the abandoned playground of the school they once attended in person. Thanks for helping me get the details straight even when the information seemed overwhelming.

To the care teams who watched over my health and my family's and so many others' during these difficult years. To Callie's teachers and coaches and dance instructors and caregivers (and Hannah!) and friends. Thank you.

Thanks to the organizations and institutions that supported me as I finished this book. The John Simon Guggenheim Memorial Foundation. The Academy of American Poets. The Sustainable Arts Foundation. Special thanks to Colorado State University for naming me a university distinguished professor and for continuing to value and work to support the humanities, creative writing, and the arts. Thank you to my colleagues, whose work I so deeply admire.

To my students at CSU and Vanderbilt and Whitman and

Stanford and all the other wonderful places where I've had the blessing to talk about the wonders of writing and the world over the past few years. Thank you for helping me keep in touch with what matters.

To the *Immaterial* team, Jesse Baker, Adwoa Gyimah-Brempong, Eleanor Kagan, Elyse Blennerhasset, Benjamin Korman, Sarah Wambold, Sofie Andersen, Rachel Smith, Eric Nuzum, Ariana Martinez, and everyone else at the Met Museum and Magnificent Noise. Though you all had nothing directly to do with this book, you helped to remind me how thrillingly wonderful it can be to work in a positive, collaborative, creative community. We made a terrific show together. I'm grateful to you for the experience of hosting the *Immaterial* podcast and for the accompanying new set of skills and knowledge.

Thanks to the publication venues in which some of the writing in this book previously appeared, often in different forms, but always with the same sincerity of purpose. I thank you for opening a place where I could begin to express some of these ideas: Terrain.org; *Dear America* (Trinity University Press); *The Georgia Review*; *The Best American Essays*; *Pushcart Prize XLIII: Best of the Small Presses 2019*; *Poetry Northwest*; NPR's *All Things Considered*; *Lit Hub*; *Guernica*; Academy of American Poets; *Emergence Magazine*; *New Daughters of Africa* (HarperCollins); *Anchor*, a magazine by Still Harbor; *All the Songs We Sing* (Carolina Wren Press); Ginkgo Prize; *The Rumpus*; *Orion* magazine; *Earthly Love* (Orion); *Sierra* magazine; *The Paris Review*; *Pacific Standard*; *Nature Swagger: Stories and Visions of Black Joy in the Outdoors* (Chronicle Books); *The American Poetry Review*; *Mothers Before* (Harry N. Abrams); *Ecotone*; *Poets & Writers* magazine; *The Atlantic*; and *The Best American Poetry*.

To Anya Backlund, Ana Paula Simões, and the whole amazing Blue Flower Arts team. Thanks for all you do to keep this show on the road.

To the best agent I could ever dream of having, Samantha Shea of Georges Borchardt, Inc. Literary Agency.

To my editor, Yahdon Israel, who kept me digging deeper and deeper, again and again. And to the rest of the team at Simon & Schuster. Lashanda Anakwah, Martha Langford, Danielle Prielipp, Sienna Farris, Imani Seymour, Natalia Olbinski, Jonathan Evans, and Ruth Lee-Mui; what a blessing.

To Lily, Bun-Bun, Bun, Sunny, Pebble, Chip, Wee-wi, and all the neighborhood bunnies.

To flowers. Now. And always.

To all my amazing neighbors, I wish I could name each of you individually. I hope you know that this appreciation is meant especially for you. Thank you for helping to make this place home. My notes of gratitude could continue for as many pages as this book. My gratitude is as abundant as a healthy garden.

Thank you. Thank you. Thank you.

Camille T. Dungy
Fort Collins, Colorado

BOOK
CLUB
FAVORITES
—
READER'S
GUIDE

Soil

The Story of

A BLACK MOTHER'S GARDEN

CAMILLE T. DUNGY

1. How does *Soil* connect the act of diversifying nonhuman spaces with that of human spaces? Why is this important?

2. What does the book's author, Camille T. Dungy, say about nature writing and its incorporation of the quotidian? Does environmental literature of the past care about the commonplace, according to Dungy? Why or why not?

3. How does Dungy interweave the personal with the political in this book? How about the material with the spiritual?

4. Analyze the key concepts in the book's title—*Soil: The Story of a Black Mother's Garden*. How and why is this book just as much about the significance of soil and gardening as it is about history and other stories we tell ourselves, about the Black American experience, and about mothering?

5. What did *Soil* teach you about John Muir (or the American West) that you didn't know before? Has your opinion of Muir (or the American West) changed after reading this book? Why or why not?

6. In what ways does this book unearth important American history, especially African American history?

7. Dungy admits that she is not perfect when it comes to environmentally conscious living. Nonetheless, what ideas does this book present for how we all can be better stewards of the Earth?

8. Has this book inspired you to begin (or revise your approach to) gardening? Why or why not?

9. What is the Great Chain of Being? In what ways does Dungy's commentary in *Soil* subvert this worldview?

10. How is *Soil* an intersectional ecofeminist text? What does Dungy say about the experiences of women (in particular, women of color) in connection with nature and the environment?

11. Why does Dungy emphasize the ability to name things in our natural world? Do you agree with her rationale? How many different kinds of native trees, plants, birds, insects, and animals can you name in the region where you live?

12. How do the cultivation practices of past African Americans and Native Americans—all the way back to the enslaved people's African origins, the Middle Passage, the slave plantations of the American South, Indigenous American's Ancestral lands, and the Trail of Tears—influence how Dungy views her present-day gardening and growing behaviors?

13. What are some of the parallels that *Soil* draws between humans and plants and/or between humans and animals? Why do you think Dungy is intent on making these comparisons?

14. Why does Dungy believe that environmental and social justice concerns are related? How does this book represent that interconnection?

15. Why does this memoir relay personal, individual stories alongside social, collective narratives? Why is so much of Dungy's book about the experiences of her family members?

16. How does *Soil* define and/or redefine these words: nature, the environment, wilderness, the wild, and a weed? Has your thinking about these terms changed at all after reading this book?